**Jesús Antonio Botello Triana**
**Juan Antonio Ramirez Bruno**
**Juana María Camarillo Escobedo**

**HMI Instrumentation in Electronic Engineering**

**Jesús Antonio Botello Triana**
**Juan Antonio Ramirez Bruno**
**Juana María Camarillo Escobedo**


# HMI Instrumentation in Electronic Engineering

## Computer monitoring of currents and voltages in three-phase electrical machines.


**ScienciaScripts**

**Imprint**
Any brand names and product names mentioned in this book are subject to trademark, brand or patent protection and are trademarks or registered trademarks of their respective holders. The use of brand names, product names, common names, trade names, product descriptions etc. even without a particular marking in this work is in no way to be construed to mean that such names may be regarded as unrestricted in respect of trademark and brand protection legislation and could thus be used by anyone.

Cover image: www.ingimage.com

This book is a translation from the original published under ISBN 978-620-2-24550-0.

Publisher:
Sciencia Scripts
is a trademark of
Dodo Books Indian Ocean Ltd. and OmniScriptum S.R.L publishing group

120 High Road, East Finchley, London, N2 9ED, United Kingdom
Str. Armeneasca 28/1, office 1, Chisinau MD-2012, Republic of Moldova, Europe
Printed at: see last page
ISBN: 978-620-5-80986-0

# INTRODUCTION

The measurement of the electrical magnitudes in an electrical machine is extremely important as it allows us to know the behaviour of the machine in many of its aspects such as voltages, currents, frequencies, etc.

Electronic Instrumentation is used to carry out these measurements. It is responsible for the design and management of electronic and electrical devices, especially for their use in measurements; it is applied in the sensing and processing of information from physical and chemical variables, from which process monitoring and control is carried out, using electronic devices and technologies.

Although the information obtained with conventional electronic instrumentation (voltmeters, ammeters, oscilloscopes, etc.) is usually sufficient in most tests and applications, the measurement of instantaneous magnitudes is necessary on many occasions, for which the combination of analogue instrumentation and virtual instrumentation can be used to show the measurements on a screen or display in a way that facilitates the analysis of the desired measurement.

One of the new trends in instrumentation is *virtual instrumentation*. The idea is to replace and extend "hardware" elements with "software" elements, using a processor (usually a PC) running a specific program, which communicates with the devices to configure them and read their measurements.

The advantages of virtual instrumentation are the ability to automate measurements, data processing, visualisation and remote actuation, among others.

Some specialised software in this field are LabVIEW and Agilent-VEE (formerly HP-VEE). And some popular communication buses are GPIB, RS-232, USB, etc.

# CHAPTER 1: PROJECT APPROACH

## Instrumentation for harmonics evaluation in electrical machines

The project starts with the assembly of the Hall effect sensors, which will allow the sensing of the electrical variables of interest in the electrical machines under test. Once the sensors are working, computer tools will be programmed using LabVIEW programming to carry out computer monitoring using a data acquisition card. To carry out this project, Hall effect current and voltage sensors will be mounted on boards with screw terminals to configure the measurement capabilities of these sensors. There is also a 9 kVA three-phase power supply, which is programmable (it can handle 1 or 2 phases) and allows the injection of harmonic currents and voltages to the loads connected to it. On the other hand, there are single-phase and three-phase rotating electrical machines, as well as a three-phase transformer.

## 1.1 JUSTIFICATION

Currently, sub-transmission and distribution level power grids in industrial areas feed industrial loads that include power electronic devices, such as variable speed drives, induction and electric arc furnaces, electric arc welding, large amounts of computer equipment, among others. This has resulted in a large part of these power grids being subjected to conducting currents that are distorted in nature, which causes what is known as harmonic pollution in power grids. This harmonic pollution has a negative impact on the operation of rotating electrical machines and electrical transformers, so it is important to investigate the problems that this causes in these machines. In view of this, it is proposed to install Hall effect sensors to monitor the currents and voltages consumed by electrical machines, in addition to developing the data acquisition platform using LabVIEW and a National Instruments data acquisition card.

## 1.2 OBJECTIVES

To install a series of Hall effect sensors for computer monitoring of currents and voltages in three-phase electrical machines in order to evaluate different aspects of the behaviour of the machines when the power supply presents harmonic contamination.

The monitoring will make it possible to visualise the variables contaminated by harmonics and, in turn, to carry out harmonic contamination analysis on the machine. The power supply to the electrical machines is provided by a programmable power supply of 9 kVAs and up to 250 VAC, which can work with one, two or three phases, applying distortion to the voltage waveform. The test stage may be performed on rotating machines that are at rest or in motion, as the source allows distortion of the nominal supply voltage for rotating machines. For transformers it is similar, but mechanical motion is not considered.

## 1.3   PROBLEMS TO BE SOLVED

Monitor harmonically contaminated voltage and current variables that are applied to an electrical machine to perform harmonic contamination analysis on the induced electromagnetic field.

## 1.4   SCOPE AND LIMITATIONS

The scope to be achieved will be the analysis of different variables in the electrical machine, the limitations will be that it will not be possible to apply high voltages or currents as we will only work with small-scale analysis to see that the sensing is being carried out correctly.

# CHAPTER 2: ENTERPRISE

**Name:** Instituto Tecnológico de La Laguna

## 2.1 BUSINESS LINE AND LOCATION OF THE COMPANY

### 2.1.1 Description of the Company and Location

The **Instituto Tecnológico de La Laguna** (ITL) is one of the most important institutions of higher technological education in the Laguna region. The Instituto Tecnológico de La Laguna is part of the National System of Technological Institutes in Mexico. Up to 1987, it provided education at the upper secondary level. In 2005, the Instituto Tecnológico de La Laguna obtained the ISO 9000:2001 accreditation, awarded by the German organisation TÜV. In April 2008 it received the distinction of High Performance Technological Institute, placing the Instituto Tecnológico de La Laguna among the 8 best institutes in the country.

**Address:** Cuauhtémoc y Revolución S/N <u>Torreón</u>, <u>Coahuila Mexico.</u>

### 2.1.2 Mission and vision of the company

**MISSION OF THE** TECNOLOGICO

To be an instrument for the integral development of society, training professionals who stand out for their innovative, creative, enterprising and humanistic skills.

**TECHNOLOGICAL VISION**

To be an institution of higher technological education that is competitive at national and international level, generating innovative knowledge to respond creatively to the challenges of globalisation, from the perspectives of equity, sustainability and the integral development of the human being.

Figure 2.1 Image of the Instituto Tecnológico de la Laguna.

## 2.2 DEPARTMENT WHERE THE RESIDENCY TOOK PLACE

## GRADUATE STUDIES AND RESEARCH DIVISION

Organigrama de la División de Estudios de Posgrado e Investigación

# CHAPTER 3: THEORETICAL BACKGROUND

This section comprises the theoretical concepts related to the project, as well as the description of the methods, techniques and tools used in the development of the project.

## 3.1 The Electric Machine

An electrical machine is a device that transforms kinetic energy into another energy, or into potential energy but with a different presentation, passing this energy through a storage stage in a magnetic field. They are classified into three main groups: generators, motors and transformers.

An electrical machine has a magnetic circuit and two electrical circuits. Normally one of the electrical circuits is called an excitation, because when it is driven by an electric current it produces the amper-turns necessary to create the flux set in the whole machine.

**Types of electrical machines**

Table 3.1 Types of rotating and static electrical machines

| Feeding | Rotating Machines | | Static machines |
|---|---|---|---|
| **Single-phase and three-phase alternating current** | Synchronous | Generator<br>Motor<br>Compensator | Transformer<br>Induction regulator<br>Phase inverter<br>Cycloconverter |
| | Asynchronous | Engine<br>Generator<br>Compensator | |
| | Switched | Single-phase motor in series<br>Frequency converter | |
| **Direct current** | Switched | Generator<br>Motor<br>Compensator | Chopper |
| **CA<br>CD** | Switched | Universal motor<br>Converter | Rectifier<br>Inverter |

### 3.1.1 Electric Motors

The electric motor is a device that transforms electrical energy into mechanical energy by means of the action of the magnetic fields generated in its coils. They are rotating electrical machines composed of a stator and a rotor.

Some of the electric motors are reversible, as they can transform mechanical energy into electrical energy by functioning as generators or dynamos. Electric traction motors used in locomotives or hybrid cars often perform both tasks, if properly designed.

Electric motors are used in the vast majority of modern machines. Their small size makes it possible to fit powerful motors into small machines, such as drills or mixers. Their high torque and high efficiency make them the ideal engine for traction in heavy transport such as trains, as well as for propulsion of ships, submarines and mining dumpers, via the diesel-electric system.

### 3.1.2 The induction motor

Asynchronous or induction motors are essentially three-phase motors. They are based on the drive of a metal mass by the action of a rotating field.

They consist of two armatures with coaxial rotating fields: one is fixed, the other is movable. They are also called, respectively, stator and rotor.

The rotor winding, which conducts the alternating current produced by induction from the directly connected stator winding, consists of copper or aluminium conductors embedded in a rotor of steel laminations. Short-circuit terminal rings are installed at both ends of the "squirrel cage" or at one end in the case of the wound rotor.

### 3.1.2.1 The Transformer Model of an Induction Motor

The equivalent circuit of an induction motor is very similar to that of a transformer, due to the transformation action that occurs when currents are induced in the rotor from the stator.

The figure below shows a transformer equivalent circuit, per phase, of an induction motor. The stator resistance is $R_1$ and the reactance is $X_1$ . $E_1$ is the primary stator voltage, coupled to the secondary $E_R$ by an ideal turns ratio transformer at$_\text{eff}$ . The voltage $E_R$ produced in the rotor causes a current flow in the rotor. $R_C$ is the core loss component and $jX_M$ refers to the magnetisation reactance, $R_R$ and $jX_R$ are the rotor impedances. Also, $I_1$ is the line current, $I_M$ is the magnetisation current. $I_R$ is the current flow in the rotor and $I_2$ is the rotor current.

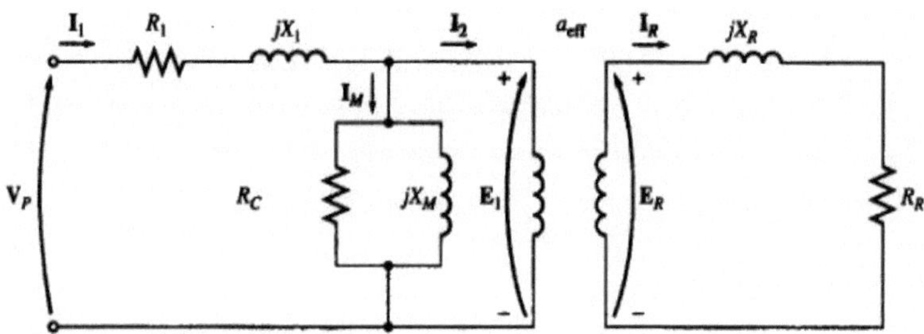


Figure 3.1 Transformer model of an induction motor.

## 3.1.2.2 Induction motor design classes.

A letter identification system has been developed for commercial engines according to the torque-speed characteristic curves for each design:

**Class A design**: A squirrel cage motor for use at constant speed. Its main characteristics are: good heat dissipation capacity, high resistance and low starting reactance, fairly fast acceleration compared to normal acceleration, starting current between 5 and 7 times the rated current, used in fans, blowers, pumps, lathes, etc.

**Class B design**: Also called general purpose motors, most squirrel cage motors belong to this type. Their main characteristics are: normal starting torque, low starting current, low slip, preferred over class A due to their low starting current consumption.

**Class C design**: These motors have high starting torque and low starting currents, are built with a double cage rotor which makes them more expensive and tend to overheat with frequent starts.

**Class D design**: Also known as high torque and high strength, high slip at full load and designed especially for heavy-duty starting.

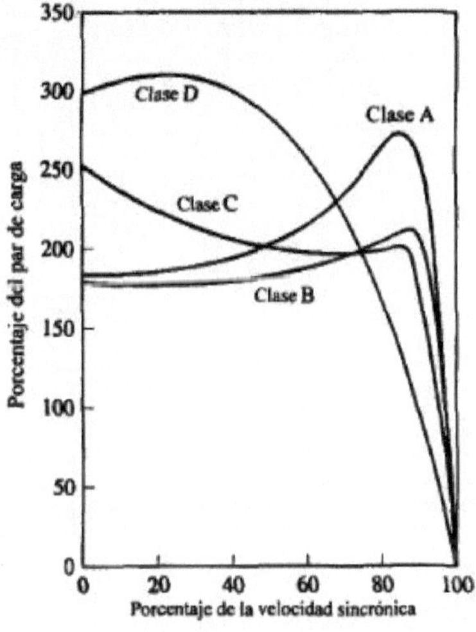


Figure 3.2 Typical characteristic curves for different engine designs.

An induction motor has the same stator as a synchronous machine but its rotor has a different construction. There are two different types of induction motor rotors, one is known as a squirrel cage rotor and the other as a wound rotor.

A squirrel cage rotor consists of a series of busbars placed inside grooves made in the rotor surface with their ends shorted by means of rings.

The so-called wound rotor, which has a complete three-phase winding that is a mirror image of the stator. The three winding phases of this rotor are usually Y-connected and their ends are connected to slip rings mounted on the shaft. The rotor windings can be short-circuited through a set of brushes that are in contact with the slip rings.

Figure 3.3 Wound rotor induction motor

### 3.1.2.3 Determination of the impedances of the equivalent circuit model.

The equivalent circuit of an induction motor is a very useful tool to determine the responses of the motor when there are variations in the load. To establish the model of the particular machine it is necessary to determine the values of the various model elements. How can the values of $R_1$ ,$R_2$ ,$X_1$ ,$X_2$ and $X_M$ of a real motor be determined?

In order to find this information, a series of tests analogous to the short-circuit and open-circuit tests carried out on transformers are carried out, and these tests are detailed below.

### Vacuum test

The vacuum test of an induction motor measures the rotational losses of the motor and provides information about its magnetising current. Three ammeters, a voltmeter and two wattmeters are connected to the input circuit and the motor is allowed to rotate freely. The only load on the motor is its friction and ventilation losses which makes this practically zero. If the input power to the motor is known by applying the input power formulae $P_{ent}$ , the rotational losses can be known, if we somehow manage to know $X_1$ , then the value of the magnetisation impedance $X_M$ of the motor can be calculated.

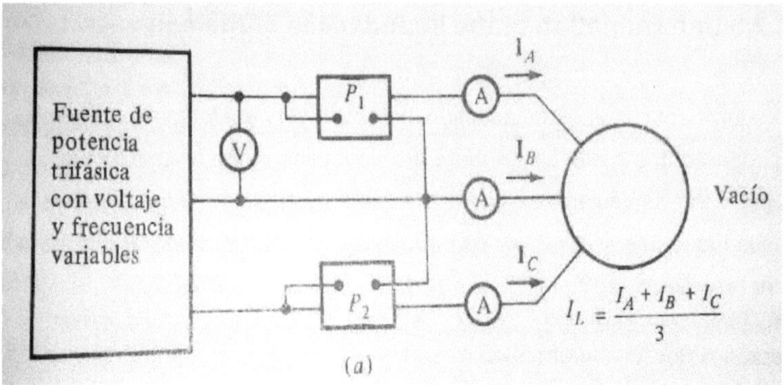


Figure 3.4 Induction motor vacuum test - Test circuit.

## Test with dc to determine stator resistance.

The rotor resistance plays an extremely critical role in the operation of an induction motor, $R_2$ determines the shape of the torque-speed curve at which the maximum output torque occurs. In order to know exactly the resistance $R_2$ it is necessary to know $R_1$ in order to subtract it from the total resistance. The test to measure $R_1$ basically consists of applying dc voltage to the stator winding terminals of the induction motor stator. By having dc current, no voltages are induced in the rotor circuit and therefore there is no current in the rotor, so the only thing that opposes the circulation of current in the motor is the stator resistance, which can be determined. During this test, the stator windings are tested for a current equal to the rated current and the voltage applied to the terminals is measured.

The current flows through both windings, so the total path resistance is 2*$R_1$ then:

$$2R_1 = \frac{Vcc}{Icc}$$

$$R_1 = \frac{Vcc}{2Icc}$$

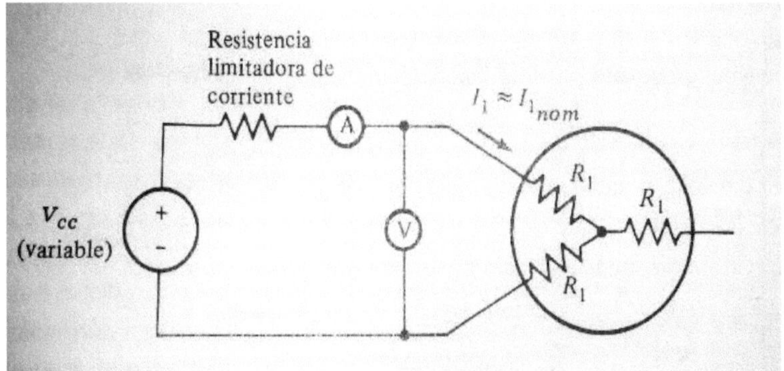


Figure 3.5 Test circuit for dc resistance measurement.

## Braked rotor test

The braked rotor test corresponds to the short-circuit test of a transformer. In this test, the rotor is braked or locked in such a way that it cannot move, voltage is applied to it and the voltage, current and power are read. The test is carried out by applying an AC voltage to the stator, of such a magnitude that it causes a current to circulate approximately equal to the nominal current, then the current, voltage and power entering the motor are measured, as the rotor does not move, the slip is s=1 and the total resistance of the rotor is $R_2$ /s is just equal to $R_2$ .

In normal operation, the stator frequency is the line frequency of the power system (50 or 60 Hz). Under starting conditions, the rotor is also at line frequency. However, under normal operating conditions, the slip of most motors is only 2 or 4% and the resulting rotor frequency is in the range of 1 to 3 Hz, which creates a problem in that the line frequency does not represent the normal operating conditions of the rotor. Since the effective rotor resistance is a function of frequency for class B and C design motors, the incorrect rotor frequency can lead to false test results. In a typical situation a frequency of 25% or less of the nominal frequency is used. While this approach is acceptable for constant resistance rotors (class A and D designs), it is not applicable when trying to find the normal rotor resistance in a variable resistance rotor. Because of these and similar problems, great care must be taken when making measurements during these tests.

After the voltage and frequency have been set for testing, the current flow in the motor is quickly adjusted close to the nominal value and the input power, voltage and current are measured before the rotor gets too hot. The input power to the motor is given by:

$$P = \sqrt{3}\, V_T\, I_L \cos\theta$$

So the locked rotor power factor can be found as follows:

$$PF = \cos\theta = \frac{Pin}{\sqrt{3}\, V_T\, I_L}$$

And the impedance angle $\theta$ is just equal to the inverse cosine PF.

The magnitude of the total impedance in the motor circuit is:

$$|Z_{LR}| = \frac{V_\theta}{I_1} = \frac{V_T}{\sqrt{3}\, I_L}$$

And the angle of the total impedance is $\theta$ accordingly,

$$Z_{LR} = R_{LR} + jX'_{LR}$$

$$= |Z_{LR}|\cos\theta + j|Z_{LR}|sen\theta$$

The locked rotor resistance R is equal to:

$$R_{LR} = R_1 + R_2$$

Whereas the locked-rotor reactance $X'_{LR}$ is equal to

$$X'_{LR} = X'_1 + X'_2$$

Where $X'_1$ y $X'_2$ are the stator and rotor reactances at the test infrequency, respectively. The rotor resistance $R_2$ can be found from:

$$R_2 = R_{LR} - R_1$$

Where $R_1$ was determined from the test the total reactance of the rotor referred to the stator can also be found. Since the reactance is directly proportional to the frequency, the total equivalent reactance at the normal operating frequency can be found as:

$$X_{LR} = \frac{f_{nominal}}{f_{test}} X'_{LR} = X_1 + X_2$$

Unfortunately, there is no simple way to separate the contributions of the reciprocal rotor and stator reactances. Over the years, experience has shown that the rotor and stator reactances are proportional in motors of certain design types. In common practice, there is no real problem with analysing $X_{LR}$ since the reactance is the sum of $X_1 + X_2$ in all torque equations.

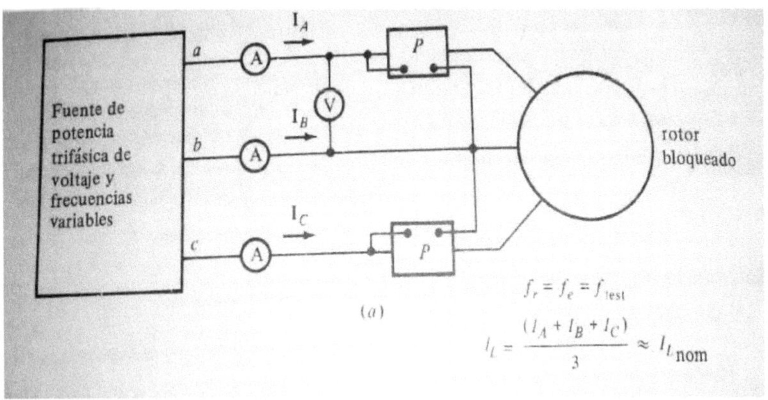


Figure 3.6 Braked rotor test of an induction motor. Test circuit.

## 3.2 The Harmonics

Harmonics are current or voltage signals, or both, present in an electrical system, which present a distorted profile with respect to the sine wave and which are composed of different frequencies that are multiples of the fundamental frequency. With the increasing use of non-linear loads (from power electronics), problems have started to occur in electrical installations, due to the generation of current and voltage harmonics in the electrical system. These include overheating of cables, transformers and motors, excessive neutral currents and resonance phenomena. Increased harmonic voltage distortion can cause incorrect operation of many equipment (especially less robust ones) that have been designed to operate under normal (low harmonic distortion) conditions.

Some equipment that generates harmonics:

- ✓ Switch Mode Power Supplies (SMPS)
- ✓ Electronic stabilisers for fluorescent lighting fixtures
- ✓ Uninterruptible Power Supply Systems (UPS)
- ✓ Electric motors.

Some problems caused by harmonics:

- ✓ Overloading of neutral conductors
- ✓ Overheating of transformers
- ✓ Untimely tripping of residual current circuit-breakers
- ✓ Overloading of the power factor compensation capacitors
- ✓ Noise and possible damage to electronic circuits
- ✓ Waveform disturbances.

Some methods to reduce harmonics:

- ✓ Passive filters
- ✓ Isolation transformers
- ✓ Active solutions

## 3.2.1 Effects of harmonics on induction motors

In power systems, motors are a very representative component of the load and are widely used in industrial and commercial installations. Induction motors are sensitive to harmonics and are subject to all power source variations, which affect their performance and operating characteristics.

## Temperature increase

The harmonic content increases the temperature in the motor. The figure below shows the temperature increase in the single-phase and three-phase induction motor as a function of the harmonic factor. It is observed that single-phase motors are more sensitive than three-phase motors. Lower order harmonics have higher effect than higher order harmonics.

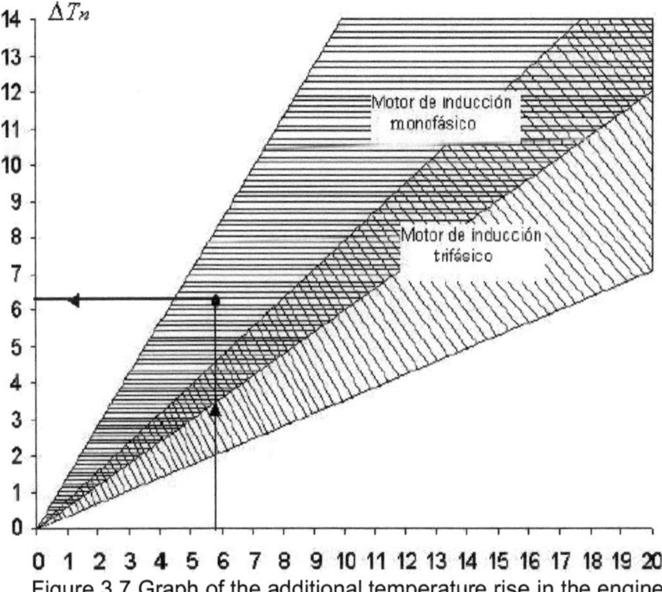


Figure 3.7 Graph of the additional temperature rise in the engine.

**Loss of capacity**

The figure below shows the capacity loss for different motor sizes, with class B insulation, considering that only the second harmonic is present. Smaller motors are more sensitive than larger ones: the higher the distortion, the higher the capacitance loss. The capacity loss is calculated with:

$$P_{p\acute{e}rdidaCapacidad} = 1 - \frac{P_{salida,n}}{P_{salida}}$$

Where: $P_{salide,n}$ is the output power of the motor when supplied from a non-sinusoidal source, $P_{salide}$ is the output power of the motor when supplied from a sinusoidal source, $P_{salide,n}/P_{salide}$ is the capacity loss factor.

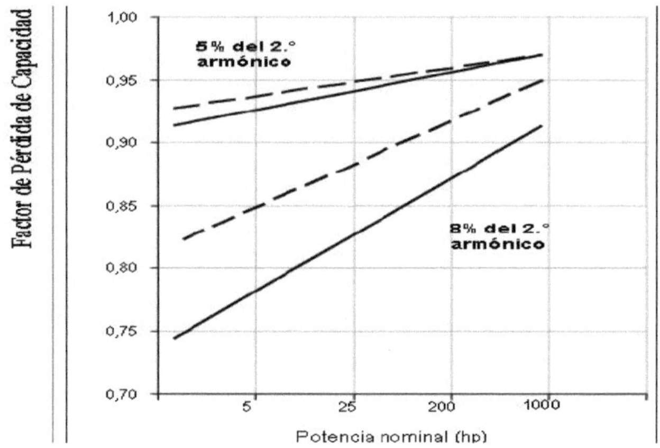


Dotted lines: radially cooled, drip-protected engine
Solid lines: fully enclosed fan-cooled engine.

Figure 3.8 Capacity loss due to harmonic voltage distortion.

## Acoustic noise and pulsating torques

Sources of acoustic noise can be divided into four categories: magnetic, mechanical, aerodynamic and electronic.

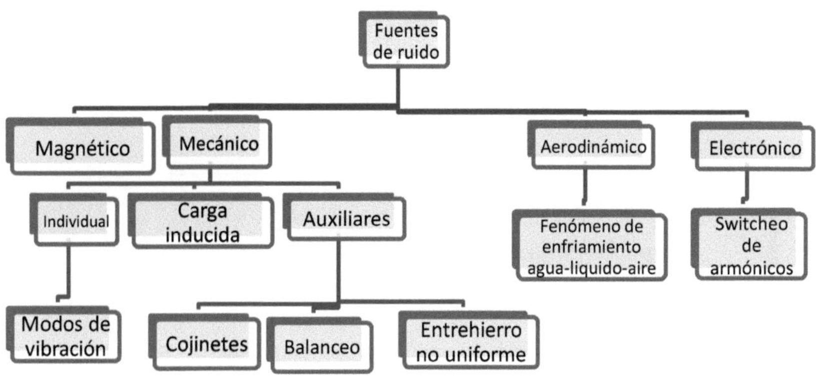


Figure 3.9 Classification of noise sources in electric motors.

In general, the contribution of harmonics to the average torque is small, approximately 4% reduction. In industrial systems, the most important harmonics are the fifth, seventh and eleventh harmonics. Generally, the electromotive forces of these harmonics are high enough to cause considerable noise and increased pulsating torques, this increase is significant if the harmonic order is close to the natural frequencies of the stator.

## Power factor and efficiency

The figure below shows the variation of the power factor as a function of individual harmonic distortion. It can be seen that the higher the voltage distortion, the lower the power factor, and that lower order harmonics have a greater effect than higher order harmonics for a given value of harmonic distortion. Lower order negative sequence harmonics (lower than fifth order) have a greater effect on power factor decrease than positive sequence and zero sequence harmonics.

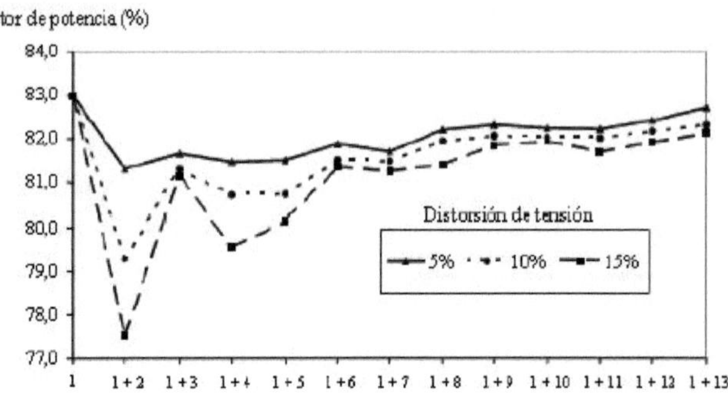


Figure 3.10 Power factor as a function of harmonic distortion.

Harmonic voltage distortion can be used as a measure of losses and temperature rise in electric motors. Motor operating conditions and lifetime are affected when harmonics are present.

Subharmonics have a dramatic effect on motors. A 6 Hz positive sequence subharmonic, for example, reduces motor life by more than 60%. Operating costs, heating and loss of lifetime resulting from electrical pollution can be minimised by decreasing harmonic voltage distortion by employing different techniques, such as: using three-phase converters with more than six pulses, using phase-shifted transformers, combining single-phase loads with three-phase loads, using line reactors at the input of the converters or using resonant filters and active filters.

Lower order harmonics have a greater effect on reducing efficiency, life, power factor and motor losses than higher order harmonics.

Lower order and negative sequence harmonics have a greater effect than positive sequence and zero sequence harmonics. Motors with a high content of voltage harmonics will suffer from increased vibration and pulsating torques.

## 3.3 The sensor

A sensor is a device capable of detecting physical or chemical quantities, called instrumentation variables, and transforming them into electrical variables. Instrumentation variables can be for example: temperature, light intensity, distance, acceleration, tilt, tilt, displacement, pressure, force, torque, humidity, motion, pH, etc. An electrical quantity can be an electrical resistance (as in an RTD), an electrical capacitance (as in a humidity sensor), an electrical voltage (as in a thermocouple), an electrical current (as in a phototransistor), etc.

Sensor application areas: Automotive industry, robotics, aerospace industry, medicine, manufacturing industry, etc.

Sensors can be connected to a computer to gain advantages such as access to a database, taking values from the sensor, etc.

**Characteristics of a sensor**

- ✓ Measuring range: domain in the measured quantity in which the sensor can be applied.
- ✓ Accuracy: is the maximum expected measurement error.
- ✓ *Offset* or zero offset: value of the output variable when the input variable is zero. If the measurement range does not reach zero values of the input variable, another reference point is usually set to define the *offset*.
- ✓ Linearity or linear correlation.
- ✓ Sensitivity of a sensor: assuming it is from input to output and the variation of the input magnitude.
- ✓ Resolution: The smallest variation of the input quantity that can be detected at the output.
- ✓ Speed of response: can be a fixed time or depend on how much the quantity to be measured varies. It depends on the ability of the system to follow the variations of the input quantity.
- ✓ Derivatives: are quantities other than the measured input quantity that influence the output variable. For example, they can be environmental

conditions such as humidity, temperature or other conditions such as ageing (oxidation, wear, etc.) of the sensor.
- ✓ Repeatability: expected error when repeating the same measurement several times.

## 3.3.1 The Hall Effect Sensor

The Hall effect occurs when a transverse magnetic field is exerted on a wire through which charges are flowing. As the magnetic force exerted on them is perpendicular to the magnetic field and its velocity (Lorentz force law), the charges are propelled towards one side of the conductor and a transverse voltage or Hall voltage ($V_H$) is generated in it. Edwin Hall (1835 - 1938) discovered the effect in 1879, which, among many other applications, helped to establish, ten years before the discovery of the electron, the fact that particles circulating in a metallic conductor have a negative charge.

The Hall-effect sensor uses the Hall-effect to measure magnetic fields or currents to determine position.

If current flows through a Hall sensor and approaches a magnetic field flowing vertically to the sensor, then the sensor creates an output voltage proportional to the product of the magnetic field strength and the current. If the value of the current is known, then the magnetic field strength can be calculated; if the magnetic field is created by current flowing through a coil or conductor, then the value of the current in the conductor or coil can be measured.

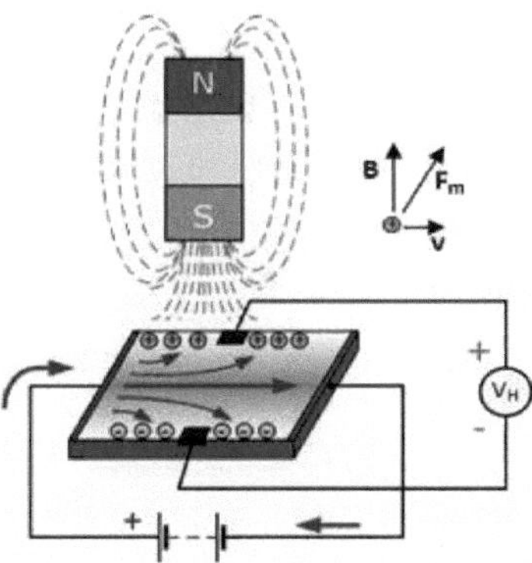

Figure 3.11 Demonstration image of the hall effect.

The picture above shows an experimental device for measuring Hall voltage. An electric current is acted on by a magnet which produces a magnetic field (**B**). The magnetic force (**F**$_m$ ) deflects the moving charges towards one side of the wire, which means that this side is charged with a charge of that sign and the opposite side is charged with a charge of the opposite sign. Consequently, an electric field and a corresponding potential difference or Hall voltage is established between the two.

## 3.4 The Fourier transform

A mathematical transformation used to transform signals between the time (or spatial) domain and the frequency domain, which has many applications in physics and engineering. It is reversible, being capable of transformations from either domain to the other. The term itself refers to both the transform operation and the function it produces.

In the case of a time-periodic function (e.g. a continuous but not necessarily sinusoidal musical sound), the Fourier transform can be simplified to calculate a discrete set of complex amplitudes, called coefficients of the Fourier series. They represent the frequency spectrum of the original time-domain signal.

The Fourier transform is basically the frequency spectrum of a function. A good example of this is what the human ear does, as it receives an auditory wave and transforms it into a decomposition into different frequencies (which is what is finally heard). The human ear perceives different frequencies as time goes by, however, the Fourier transform contains all the frequencies of the time during which the signal existed; that is, in the Fourier transform we obtain a single frequency spectrum for the whole function.

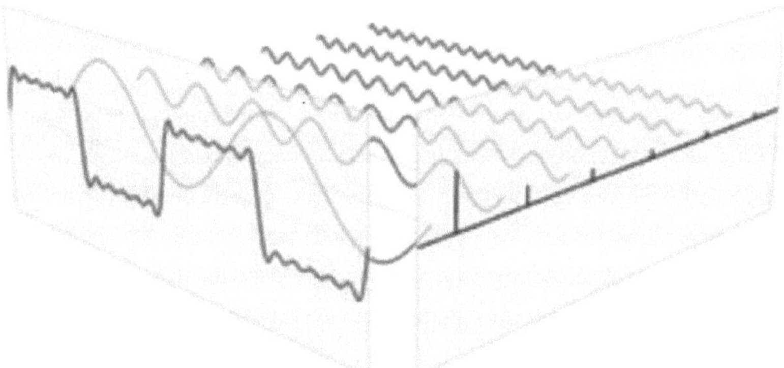

Figure 3.12 Explanation of the Fourier transform.

The Fourier transform relates a function in the time domain, shown in red in the figure above, to a function in the frequency domain, shown in blue. The component frequencies, extended over the entire frequency spectrum, are represented as peaks in the frequency domain.

# CHAPTER 4: PROCEDURE OF THE ACTIVITIES CARRIED OUT

This chapter describes how each of the activities necessary for the realisation of the project was carried out.

## 4.1 Survey of Hall-effect sensors and wiring diagrams

This section discusses hall-effect current and voltage sensors, detailing characteristics of each, and wiring diagrams.

### 4.1.2 Current Sensor

**LEM LTS 25-NP**

Used for electronic current measurement: DC, AC, pulsed, mixed with galvanic isolation between the primary circuit (high power) and the secondary circuit (electronic circuit).

Characteristics:
- ✓ Closed-loop (compensated) multirange current transducer using the Hall effect
- ✓ Single-pole voltage supply
- ✓ Compact design for PCB (printed circuit board) mounting
- ✓ Extended measuring range

Advantages:
- ✓ Excellent accuracy
- ✓ Very good linearity
- ✓ Very low temperature drift
- ✓ Optimised response time
- ✓ Wide frequency bandwidth
- ✓ No insertion loss
- ✓ High immunity to external interference

Applications:

- ✓ Variable speed controllers and servo motor drives.
- ✓ Static converters of DC motor drives.
- ✓ Supplied battery applications.
- ✓ Uninterruptible Power Supplies (UPS).
- ✓ Switch Mode Power Supplies (SMPS)
- ✓ Power supplies for welding applications.

Domain of application
:

- ✓ Industrial.

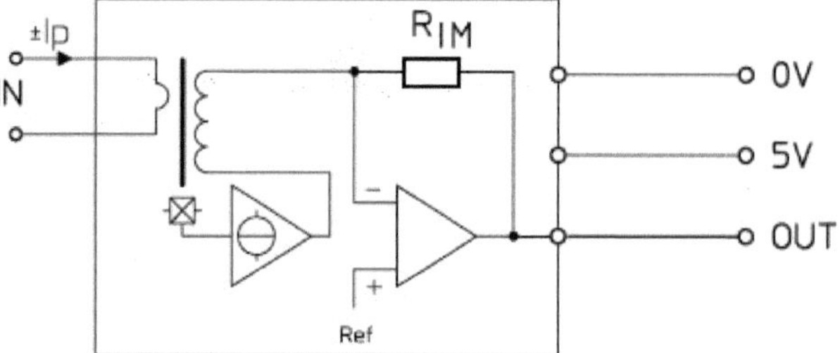


Figure 4.1 Main operation of the LTS 25-NP current sensor.

Table 4.1 LTS 25-NP sensor configuration according to current capability

| Number of primary turns | Primary nominal current rms $I_{PN}$ [A] | Nominal output voltage $V_{OUT}$ [V] | Primary resistance $R_P$ [mΩ] | Primary insertion inductance $L_P$ [µH] | Recommended connections |
|---|---|---|---|---|---|
| 1 | ± 25 | 2.5 ± 0.625 | 0.18 | 0.013 | |
| 2 | ± 12 | 2.5 ± 0.600 | 0.81 | 0.05 | |
| 3 | ± 8 | 2.5 ± 0.600 | 1.62 | 0.12 | |

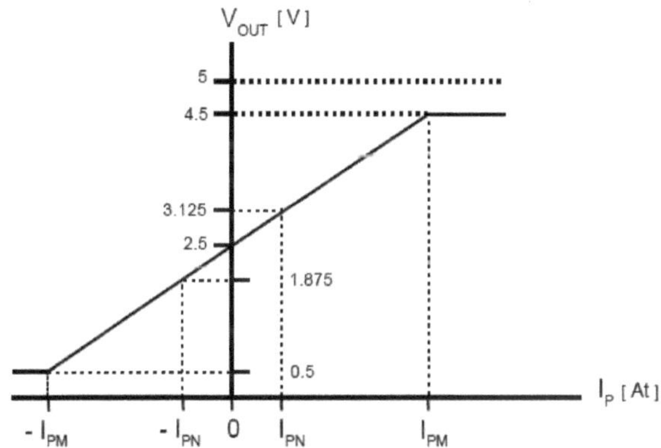


Figure 4.2 Characteristic curve of the LTS 25-NP current sensor.

The source used to power the LEM LTS 25-NP current sensors was a TEKTRONIX model PS280 Variable DC power supply, using the output that delivers 5V as fixed voltage and 3 A as maximum current.

Figure 4.3 Power supply used to power the current sensors.

### 4.1.3 Voltage Sensor

**LEM LV 25-P**

This sensor is used for the electronic measurement of voltages: DC, AC, pulses, etc., with galvanic isolation between the primary circuit (high voltage) and the secondary circuit (electronic circuit).

IPN = 10 mA

NPV = 10 . 500 V

Characteristics:

✓ Closed-loop (compensated) voltage transducer using the Hall effect.

✓ Plastic enclosure with recognised insulation according to UL 94-V0.

Principle of use:

✓ For voltage measurements, a current proportional to the measured voltage must be passed through an external resistor R1 which is user-selected and installed in series with the primary circuit of the transducer.

Advantages:

✓ Excellent precision.

✓ Very good linearity.

✓ Low thermal drift.

  ✓  High bandwidth.

  ✓  High immunity to external interference

  ✓  Low common mode disturbance.

Applications:

  ✓  Variable controllers and servo motor drives.

  ✓  Static converters of DC motor drives.

  ✓  Applications Battery supplied.

  ✓  Uninterruptible Power Supply (UPS).

  ✓  Power supplies for welding applications.


Secondary terminals

Terminal +: Supply voltage + 12 ... 15 V.

M terminal: Measurement.

Terminal -: Supply voltage - 12 ... 15 V.

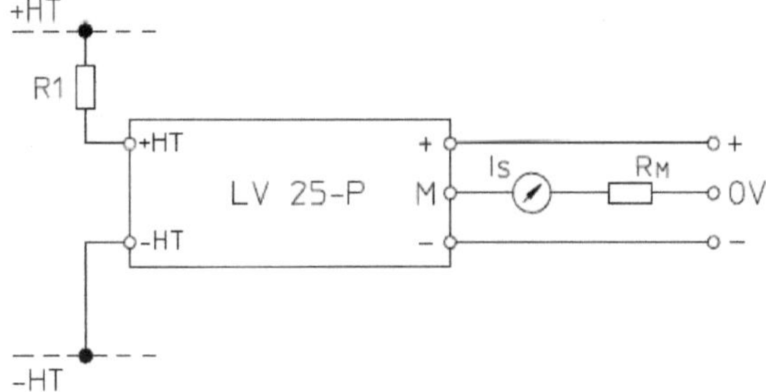


Figure 4.4 LEM LV 25-P voltage sensor connection.


The source used to power the LV 25-P voltage sensors was a 120 V to 36 V centre-tapped transformer with MC 7815 (positive voltage regulator +15V) and MC7915 (negative voltage regulator -15V) voltage regulators to make this a bipolar source.

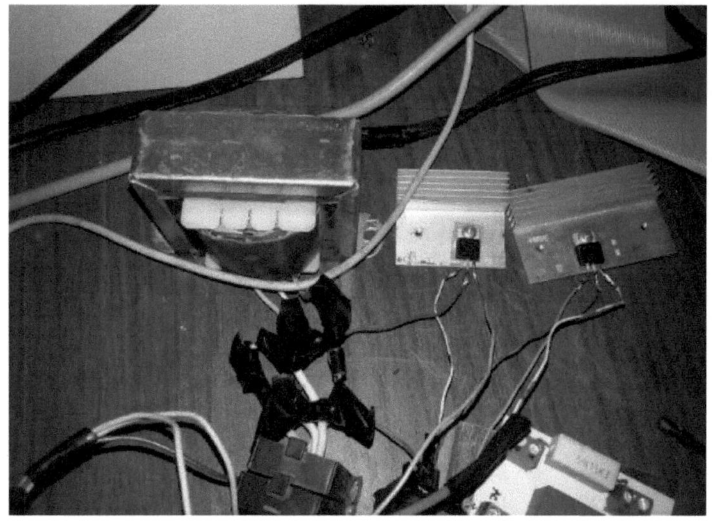

Figure 4.5 Transformer used as a bipolar source.

## 4.2 Assembly of boards and circuits for Hall Effect sensors

The PROTEUS DESIGN SUITE Version 8.1 program was used to create the diagrams for the sensors. This program was used to create the diagrams that were used to assemble the Hall Effect sensor boards (current and voltage). In this assembly of plates, its purpose is to have the hall effect sensors fixed, as well as their respective inputs and outputs to connectors which will facilitate and make the connection and disconnection of power, inputs, outputs and measurements in an orderly manner.

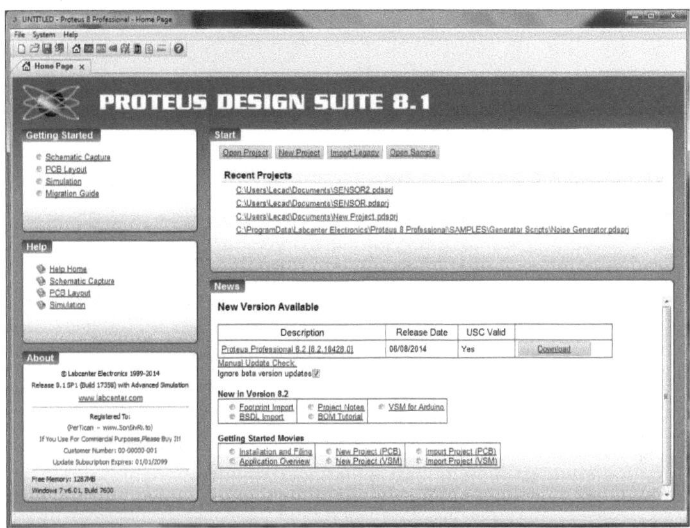

Figure 4.6 Program used to make diagrams to assemble the boards.

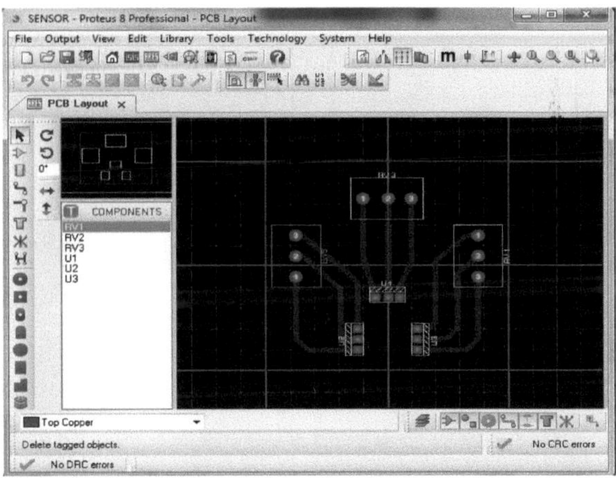

Figure 4.7 Diagram for LTS 25-NP Current Sensor in PROTEUS DESIGN SUITE.

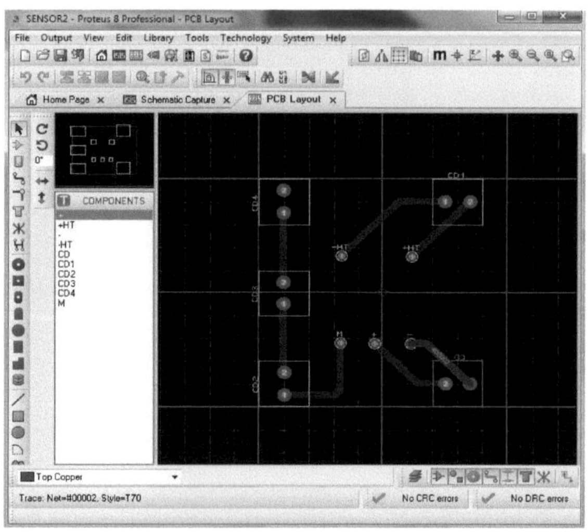

Figure 4.8 Diagram for LV 25-P Voltage Sensor in PROTEUS DESIGN SUITE.

Figure 4.9 LTS 25-NP Current Sensor mounted on phenolic plate.

Figure 4.10 LV 25-P Voltage Sensor mounted on phenolic plate.

## 4.3 Learning the LabVIEW computer language

LabVIEW is a graphical programming platform that helps engineers scale from design to test and from small systems to large systems. It offers unprecedented integration with existing legacy software, IP, and hardware by leveraging the latest computing technologies. LabVIEW provides tools to solve today's problems and the capacity for future innovation, faster and more efficiently.

LabVIEW software is ideal for any measurement and control system and the heart of the NI design platform. By integrating all the tools engineers and scientists need to build a wide variety of applications in much less time, NI LabVIEW is a development environment for problem solving, accelerated productivity, and constant innovation.

It features facilities for the handling of:

- Communications interfaces:
    - Serial port

- o Parallel port
  - o GPIB
  - o PXI
  - o VXI
  - o TCP/IP, UDP, DataSocket
  - o Irda
  - o Bluetooth
  - o USB
  - o OPC
- Ability to interact with other languages and applications:
  - o DLL: function libraries
  - o .NET
  - o ActiveX
  - o Multisim
  - o Matlab/Simulink
  - o AutoCAD, SolidWorks, etc.
- Graphical and textual tools for digital signal processing.
- Visualisation and handling of graphs with dynamic data.
- Motion control (even combined with all of the above).
- Real Time strictly speaking.
- Programming of FPGAs for control or validation.
- Synchronisation between devices.

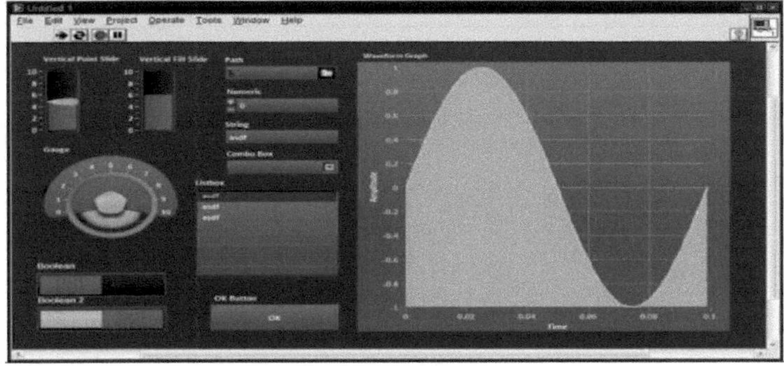

Figure 4.11 Example of a graphical program created in LabVIEW

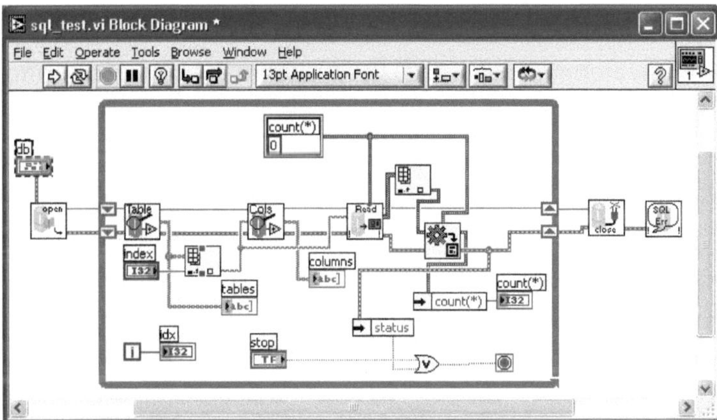

Figure 4.12 Block diagram of the example programmed in LabVIEW.

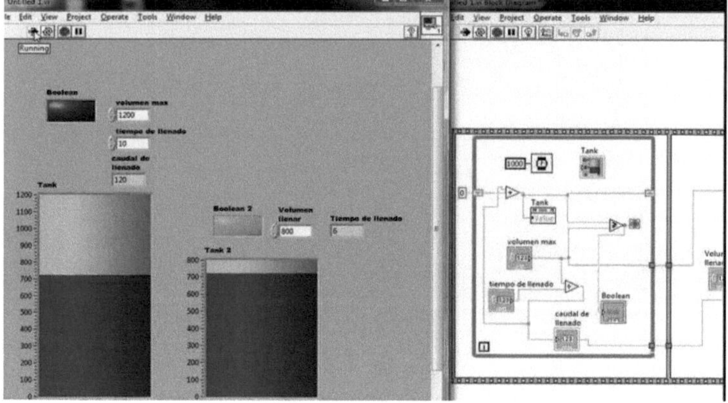

Figure 4.13 LabVIEW simulating an "X" program.

## 4.4 Development of virtual instruments for the monitoring of Hall effect sensors.

This system is based on *virtual instrumentation*, consisting of using a computer and specific *hardware* as a measuring instrument, where the screen emulates the front panel of a conventional measuring device, a set of these, or a new measuring device. The *hardware* allows the acquisition of analogue voltage signals from various channels, attenuating and adapting their value for subsequent transmission to the computer.

The application, designed using the LabVIEW programming language, allows the data acquired to be processed and displayed on the screen, thus replacing the physical components of conventional measuring equipment.

The measuring equipment used was as follows:

- ✓ PC Pentium 4 1.3 GHz - 1024 Mb RAM.
- ✓ NI PCI-6025e data acquisition board

- ✓ 8 LEM LTS 25-NP hall effect current sensors mounted on a phenolic plate with their respective inputs and outputs.
- ✓ 4 LEM LV 25-P hall effect voltage sensors mounted on a phenolic plate with their respective inputs and outputs.
- ✓ AGILENT model E3632A variable voltage source that delivers a variable voltage from 0 to 15 V with a maximum current of 7 A and a variable voltage from 0 to 30 V with a maximum current of 4 A.

### 4.4.1 NI PCI-6025e data acquisition board

16 Channels, 200 kS/s, 12 Bits, 2 AO, 32 DIO, 2 24 Bit Counters

The QUARC driver name for this card is ni_pci_6025e.

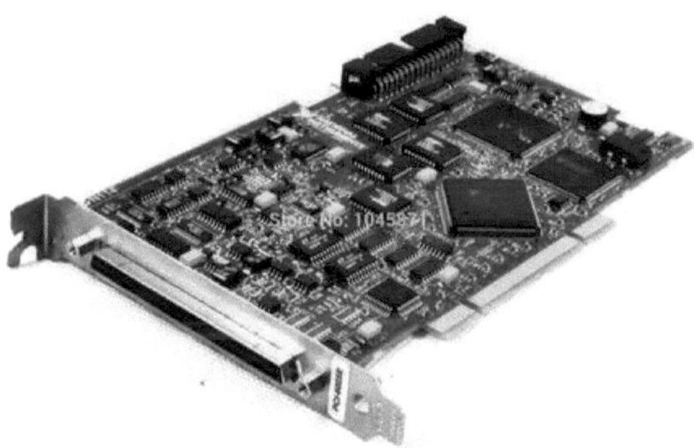

Figure 4.14 Image NI PCI-6025e data acquisition board.

**Analogue inputs**

The PCI-6025e NI supports 16 analogue inputs (channels 0-15) with 12-bit resolution. Valid input ranges are 10V, 5V, 0.5V and 0.05V. To change from the default range of 10V, the selected channels are set in the "Analogue Inputs" tab of the HIL block, "Initialise Channels". For example, the term 4:7 indicates the monitoring of channels 4 to 7. Specify [0, 4, 5] to indicate channels 0, 4 and 5.

**Digital inputs**

The PCI-6025e NI supports 32 digital input lines (P0.0 to P3.7 lines). A digital I/O line cannot be used as an input and output at the same time.

Since the digital I/O lines can be individually programmed as inputs or outputs, all channels to be used for digital inputs must be configured in the "HIL Digital Inputs" tab, "Initialise Block". Set the digital input channels field to all digital I/O channels to be used as digital inputs on the board for the current diagram. For example, enter 2: 5 to designate lines 2 through 5 as digital inputs. Specify [0, 4, 5] to indicate that lines 0, 4, and 5 are to be configured as digital inputs.

**Digital Outputs**

The PCI-6025e NI supports 32 digital output lines (P0.0 to P3.7 lines). A digital I/O line cannot be used as an input and output at the same time.

Since the digital I/O lines can be individually programmed as inputs or outputs, all channels to be used for digital outputs must be configured in the "HIL Digital Outputs" tab, "Initialise Block". Set the digital output channels field to all digital I/O channels to be used as digital outputs on the board for the current diagram. For example, enter 2: 5 to designate lines 2 to 5 outputs as digital. Specify [0, 4, 5] to indicate that lines 0, 4 and 5 are to be configured as digital outputs.

To set the digital output values when the model is loaded or unloaded, set the initial digital outputs and final digital outputs to the desired values, respectively. If the vectors specified in these fields are shorter than the channel vector, then the value of the last element in the vector is used for the rest of the channels. Therefore, a scalar value shall be applied to all channels specified in the digital output channels field.

## Encoder inputs

The PCI-6025e NI supports 2 encoder inputs with 24-bit computational values (channels 0 and 1). Note that since this board uses its counters for both encoder inputs and PWM outputs, only one counter can be used as an encoder input or PWM output channel.

In order to set the encoder counters to a particular count, when the model is loaded, the encoder inputs must be configured in the "HIL block encoder inputs" tab, "Initialise". Set the encoder input channels field to all encoder channels to be used on the board for the current diagram. The PCI-6025e NI supports only non-quadrature (count and direction) counting. Also, the PCI-6025e NI does not support encoder filtering.

## PWM outputs

The PCI-6025e NI supports 2 24-bit PWM outputs using its 2 counters (channels 0 and 1). The PWM outputs on this board use a 20 MHz counter time base rate. Note that since this card uses its counters for both encoder inputs and PWM outputs, only one counter can be used as an encoder input or PWM output channel.

To configure the PWM mode or frequency, or to set the value of the PWM outputs when the model is loaded or unloaded, the PWM outputs must be configured in the PWM outputs tab of the Initialise block HIL. Set the PWM output channels field for all PWM output channels to be used on the board for the current diagram. For example, enter 0: 1 to indicate channels 0 and 1. Specify 1 to indicate channel

1 only. In addition, the PWM output SET parameters in the model start box should always be checked.

## 4.4.2 Programming of virtual instruments in LabVIEW for the monitoring of hall effect sensors.

For the realisation of virtual instruments to be used in the monitoring of the Hall effect sensors, the first step is to connect the data acquisition card (NI PCI-6025e) to the computer and install the drivers from National Instruments so that the card can be detected.

We proceed to make the graphical program in NI LabVIEW, which, next the steps for the realisation of the program are detailed:

1.- Add the DAQ Assistant block which will extract the information from the data acquisition card, in this DAQ Assistant block the analogue input channels are registered, 12 channels will be registered, due to the fact that there will be 12 sensors that will provide information about the measurements.

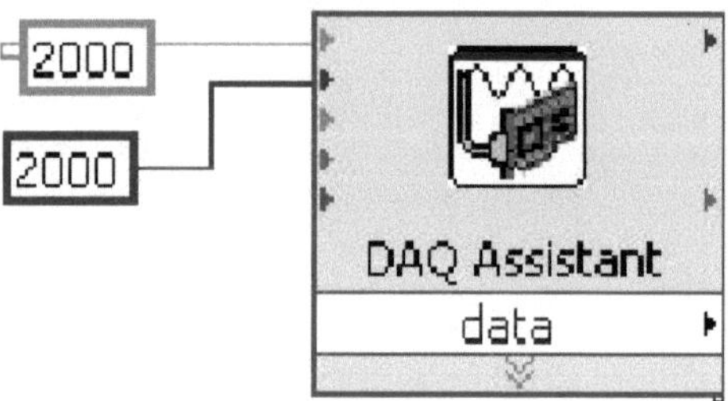



Figure 4.15 DAQ Assistant block in LabVIEW.

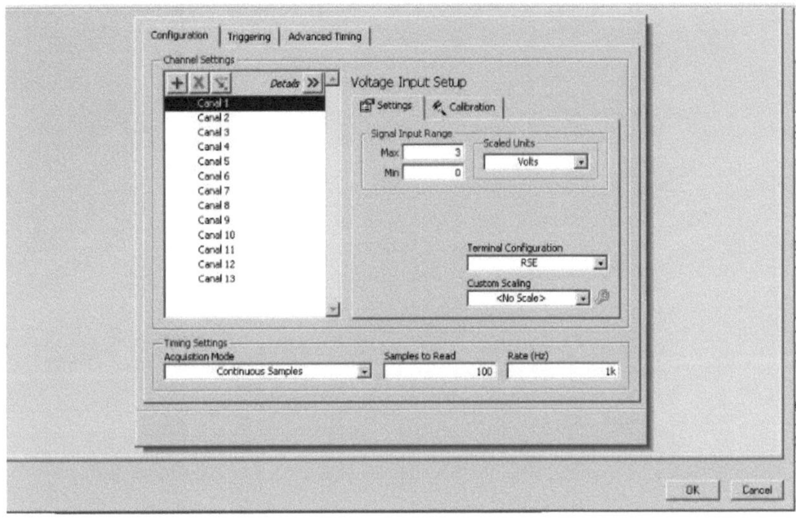

Figure 4.16 Configuration for adding channels in the DAQ Assistant block.


2.- In LabVIEW a block called Select Signals is added, from the data output of the DAQ Assistant block is connected to the signals input of the Select Signals block, in this block the desired signal is selected, this will be done 12 times since there are 12 different signals coming from the hall effect sensors of voltages and currents.

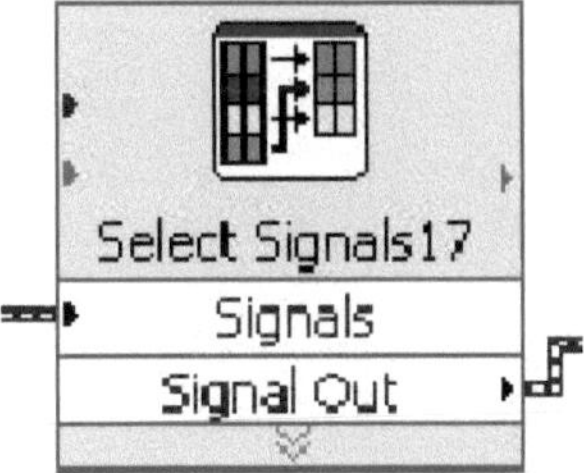


Figure 4.17 LabVIEW block to select signals.

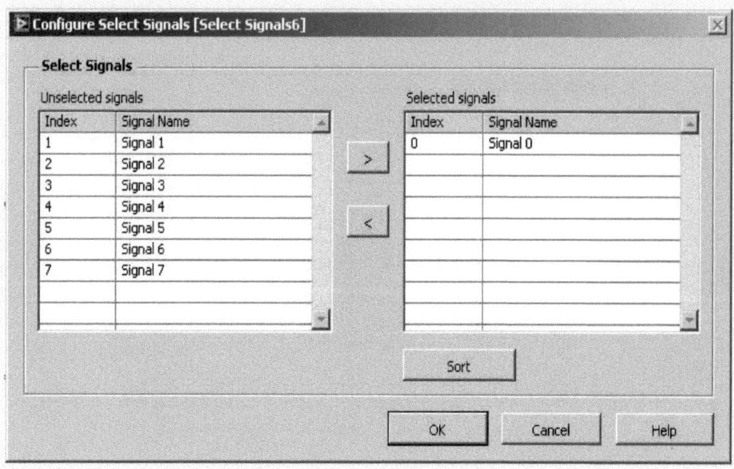

Figure 4.18 Configuration for signal selection.

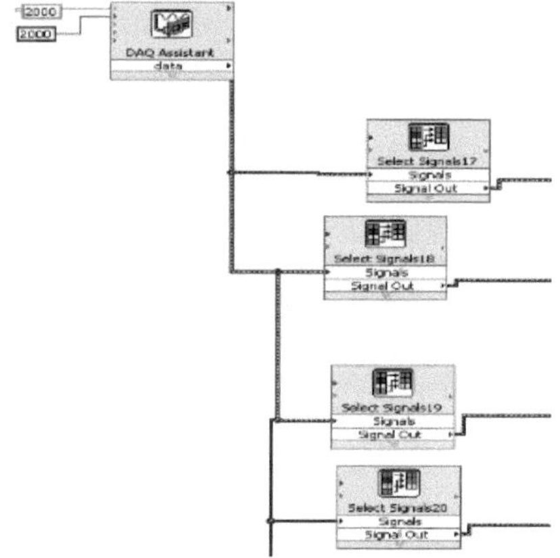

Figure 4.19 Select Signals blocks connected to DAQ Assistant

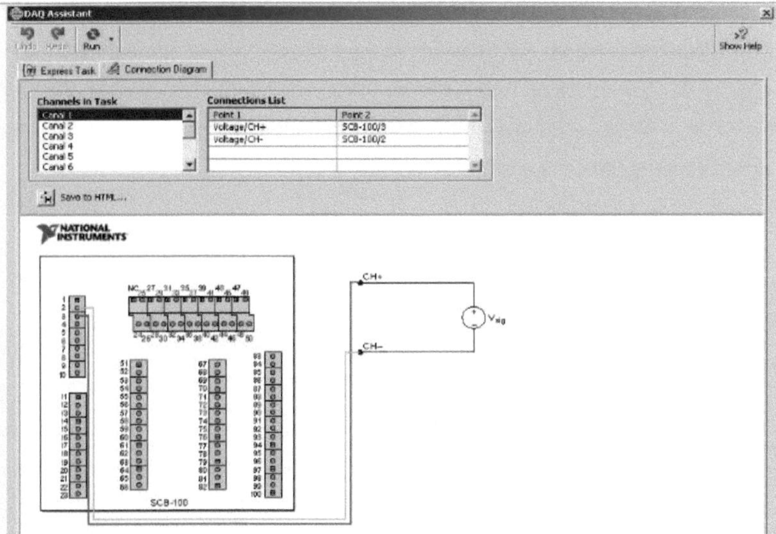

Figure 4.20 Physical connection diagram for channels.

3.- Once the signal extracted from the DAQ Assistant block has been selected, a constant is connected to the output of this block that will subtract the value and then another constant that will be multiplied so that at the end it shows the final value in the case of the current sensors. In the case of voltage sensors, only one value calculated for each sensor will be multiplied.

In the case of the current sensor signal, due to their characteristic curve (figure 4.2), they deliver around 2.5 V at their output when a current of 0 Amps circulates, but this output voltage varies for each sensor. In order to eliminate this offset in the measurement of the sensor, we proceeded to sample data at its output to calculate the arithmetic mean using a LabVIEW block called Statistics, this block gives the arithmetic mean of the signals that are introduced to the block. This step is performed for each of the current sensors.

By eliminating the offset of the sensor output, the next step is to multiply the result by a calculated constant (12.8), which in the end will give an approximate actual value of the current being measured by the respective current sensor. To view this value, a block called Waveform Graph was added to graphically display the current value at the sensor of the selected signal.

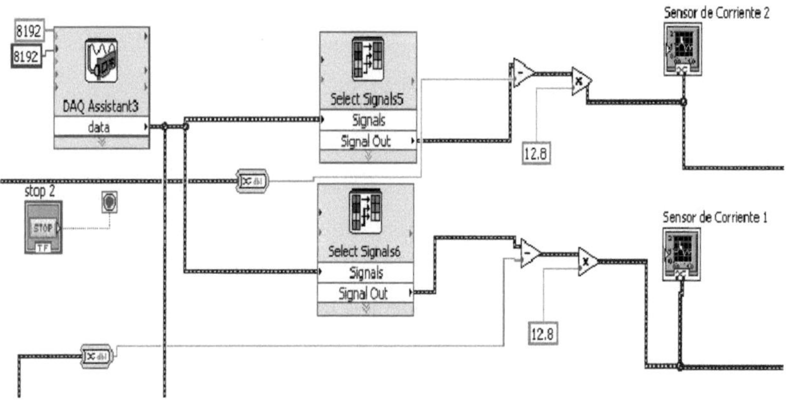


Figure 4.21 Block programming for offset elimination of the current sensors

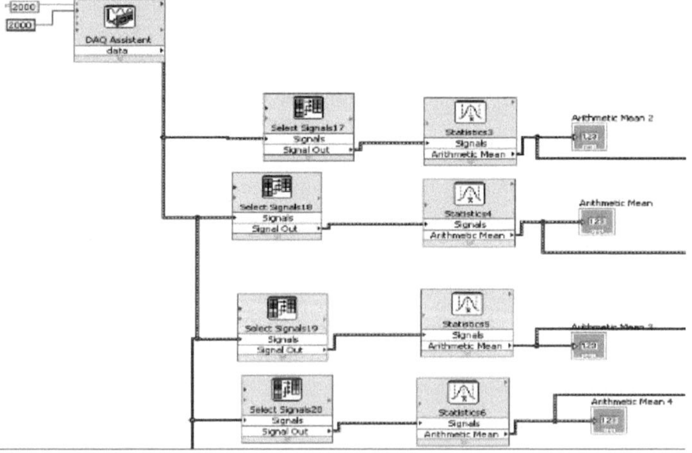


Figure 4.22 Block programming for obtaining the arithmetic mean.

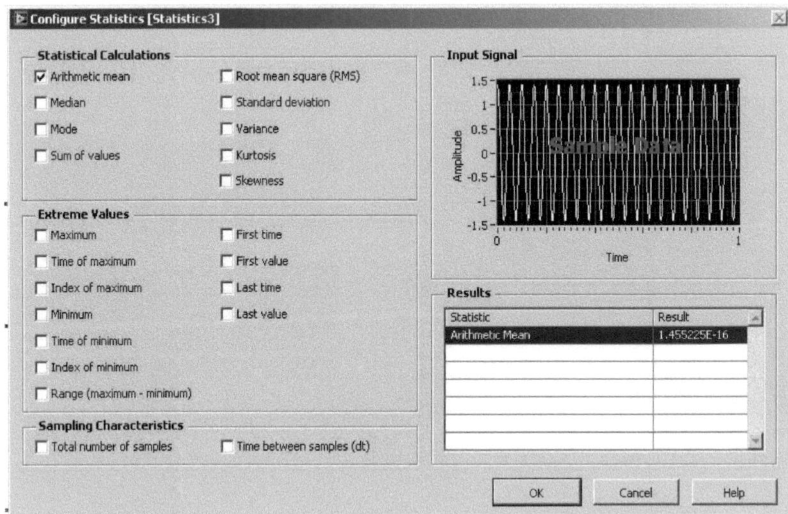


Figure 4.23 Configuration for the selection of the arithmetic mean.

For voltage sensors, the graphical programming is simpler than for current sensors, as only a calculated constant is added to the output of the signal selector depending on each sensor. This constant was calculated by trial and error using a multimeter and a power supply to feed the sensors as a reference. A block called Waveform Graph was added to the output of the multiplying constant to graphically display the voltage value at the sensor of the selected signal; this is also done for the remaining three voltage sensors.

4.- Having programmed each channel with its respective graph in blocks, all of the above is introduced in a while cycle, so that the program continues to run and thus read the voltage and current variables.

5.- Finally, we proceed to carry out the graphic measurements on the hall effect current and voltage sensors. An AGILENT model E3632A variable voltage source was used to induce voltages and currents in voltage and current sensors for measurement; the measurements are shown below.

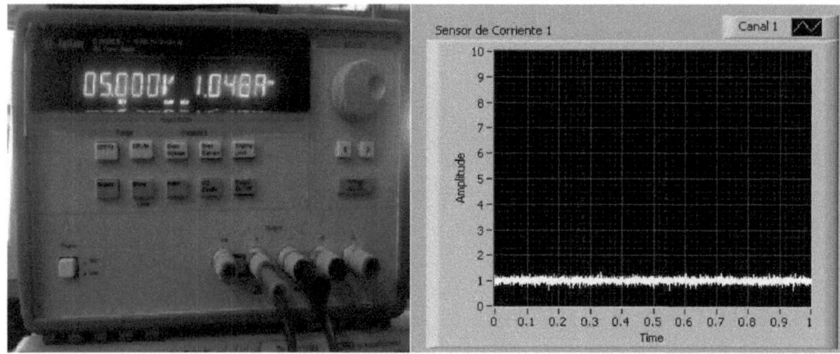


Figure 4.24 Current at 1,048 A at the source and its measurement in LabVIEW.

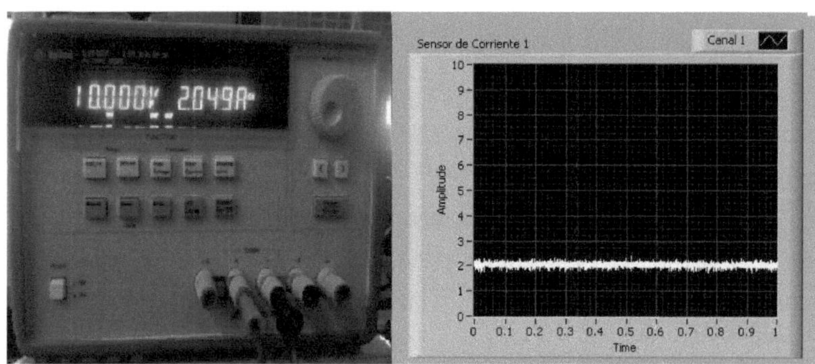


Figure 4.25 Current at 2,049 A at the source and its measurement in LabVIEW.

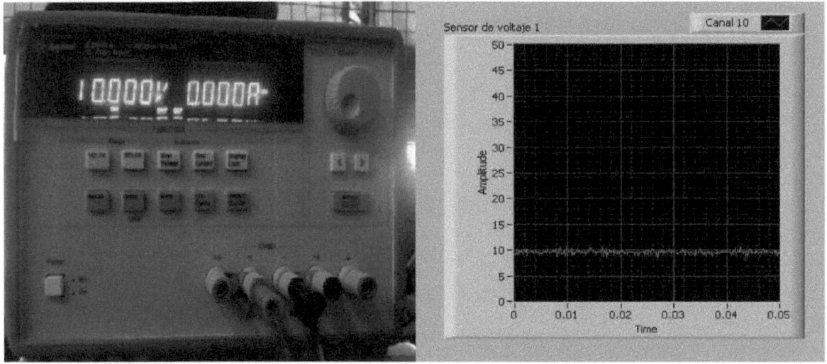

Figure 4.26 Voltage at 10 V at the source and its measurement in LabVIEW....

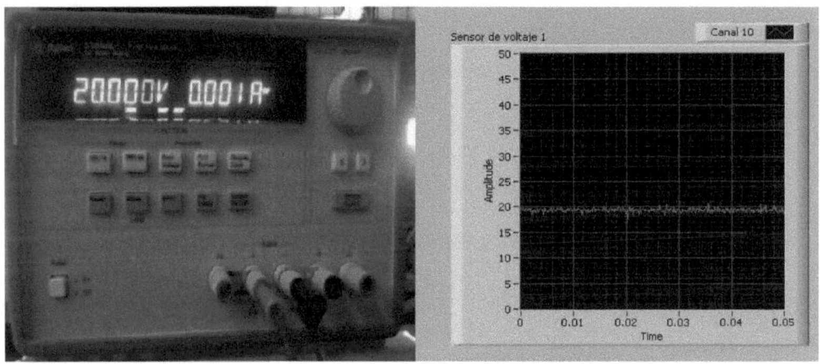

Figure 4.27 Voltage at 20 V at the source and its measurement in LabVIEW

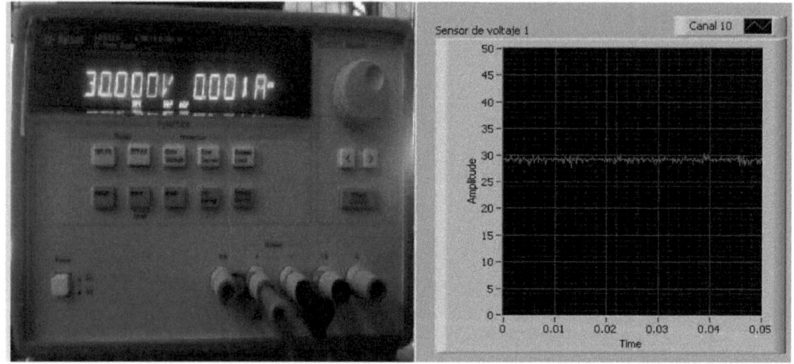

Figure 4.28 Voltage at 30 V at the source and its measurement in LabVIEW....

The above tests were performed on 8 current sensors and 4 voltage sensors and the results were found to agree with the values entered into them, therefore, this concludes the development of virtual instruments in LabVIEW for monitoring Hall effect voltage and current sensors.

## 4.5 Instrumentation of a three-phase electrical machine

The three-phase electrical machine used for testing and instrumentation was a wound rotor induction motor with the following characteristics:

Brand: Ing. de Lorenzo & C
Model: ULE500M
Rated power: 3.5 Kw
Rated speed: 1700 RPM
Frequency: 60 HZ
Supply voltage: 220/440 V
Rated current: 13.47 A
Secondary open-circuit voltage: 222 V

Figure 4.29 Three-phase wound rotor induction motor used Model: ULE500M.

The instrumentation of the electrical machine consisted of placing 9 hall effect sensors on the electrical machine, 3 current sensors to measure the current in the stator, 3 current sensors to measure the current in the rotor and 3 voltage sensors to measure the voltages in the stator as well as the respective frequency and amplitude of these voltages.

Each of the source outputs will go to a current sensor and the output of the current sensor will go to the corresponding phase in the stator, all this to measure the currents in the stator.

Each of the source outputs will enter a voltage sensor at +HT and the -HT terminals of the sensors will be put in star arrangement, all this to measure the voltages in the stator as well as to know the frequency and magnitudes in the electrical machine.

Each of the terminals of the rotor circuit will enter a current sensor and the output of these sensors will be shorted, all this to measure the currents in the rotor.

Once the connections had been made in the electrical machine, the sensor measurement outputs were connected to a connector on the data acquisition card for subsequent reading of voltages, currents and frequencies in a graphic programme in NI LabVIEW.

Finally, the electrical machine is ready to be connected to a three-phase source, which will be detailed below, as well as its connections and configurations.

Figure 4.30 Installation of voltage and current sensors on a panel.

Figure 4.31 Sensor connection board.

Figure 4.32 Data acquisition board connector.

## 4.6 Commissioning of three-phase power supply with Harmonic injection

The three-phase power supply used in the project was a *Chroma* model 6590 Programmable AC Power Supply, which handles 9 KVA, a voltage range of 0 to 150 V with a maximum current of 30 A and a voltage range of 0 to 300 V with a maximum current of 15 A RMS and a variable frequency of 45 Hz to 1Khz. The source is programmable, so the variables of its outputs can be modified and applied to the electrical machine, for example: it can work with different phases, voltages, waveforms and variable frequencies, and also allows the injection of harmonic currents and voltages to the load that is connected to it. Its characteristics are detailed in the Annex.

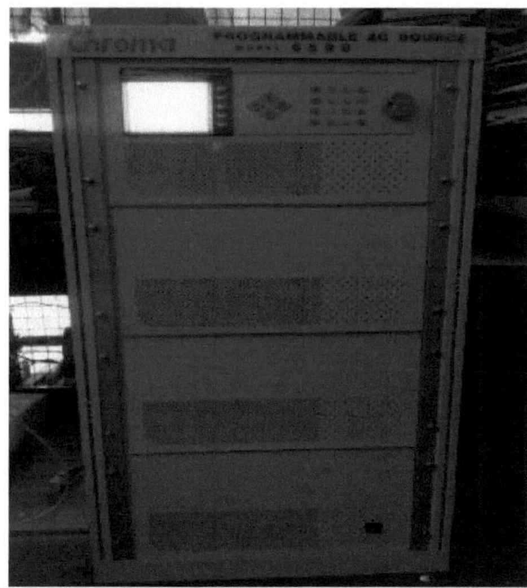

Figure 4.33 *Chroma* model 6590 three-phase programmable AC programmable power supply used in the project.

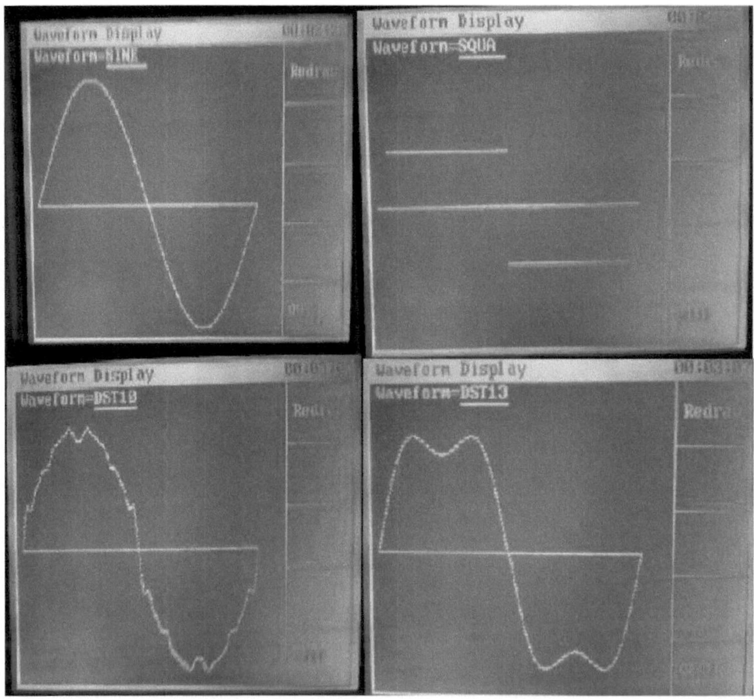

Figure 4.34 Source programmed with different output waveforms including harmonics.

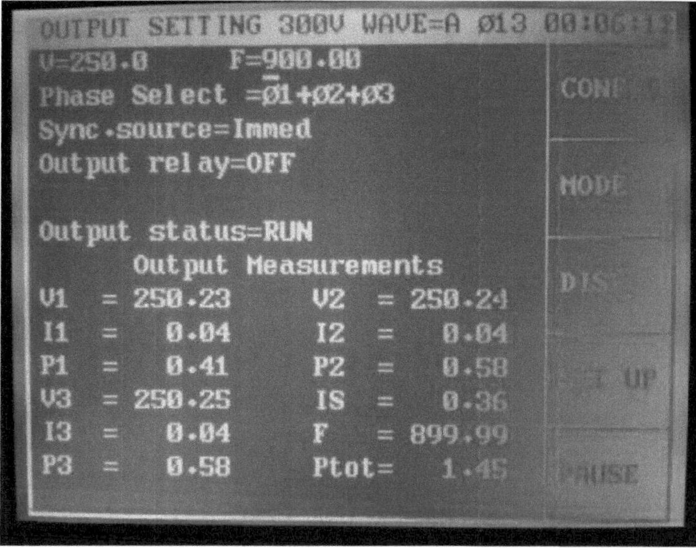

Figure 4.35 Source programmed at 150 V, 60 Hz. and 3 phases.

Figure 4.36 AC source programmed at 250 V, 900 Hz and 3 phases.

## 4.7 Development of virtual instruments for analysis on the electrical machine.

In this section, virtual instruments will be developed in LabVIEW that will be used for the analysis of the electrical machine, such as voltages, currents, frequencies, magnitudes in both rotor and stator, in order to carry out a detailed analysis of the behaviour.

### 4.7.1 Programming in LabVIEW

The programming in LabVIEW starts by adding the DAQ Assistant block which will extract the information from the data acquisition card, in this DAQ Assistant block the analogue input channels will be added, 9 channels will be added, because there will be 9 sensors (6 current sensors and 3 voltage sensors) that will provide information about the measurements.

Then 9 blocks called Select Signals are added, from the data output of the DAQ Assistant block is connected to the input signals of the Select Signals blocks, in each block the desired signal is selected, this will be done 9 times since there are 9 different signals coming from the hall effect sensors of voltages and currents.

The steps that were carried out previously when testing the virtual instrumentation of the sensors are repeated so that the LabVIEW graphs deliver a real measurement.

Continuing with the realisation of the program, 9 Waveform Graph blocks are added to the output of the selected signals, the first three blocks will show the current measured in the stator of the electrical machine, the next three blocks will show the voltages in each stator coil, the next three blocks will show the currents in the rotor.

Three blocks called Measurement Spectral were added to the outputs of the selected voltage signals and to the output of each of these a Waveform Graph block was added to observe the measurement spectra and show the frequency measurement as well as the magnitude.

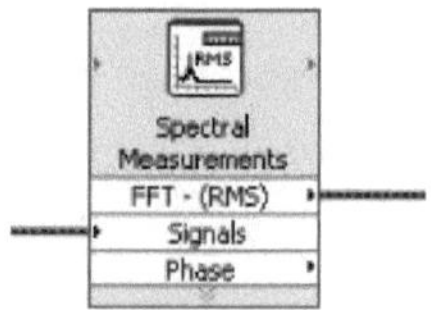

Figure 4.37 LabVIEW Measurement Spectral block.

At the end of the program, all the blocks are introduced in a while loop, so that the program continues executing and reads the variables of voltages, currents, frequencies and magnitudes.

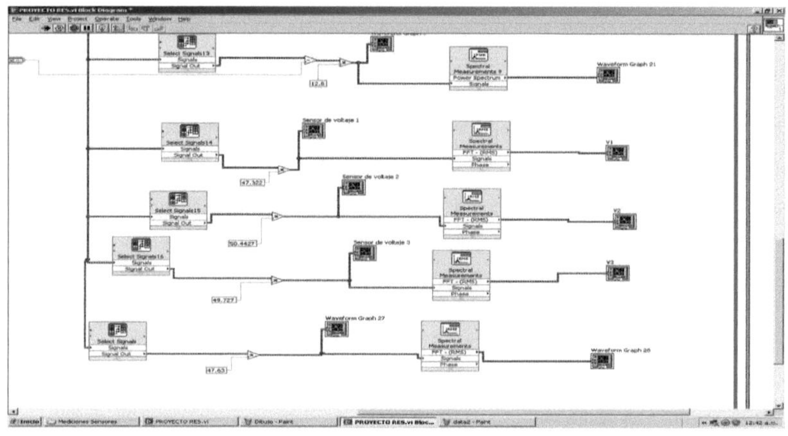

Figure 4.38 Image 1 of the program created in LabVIEW.

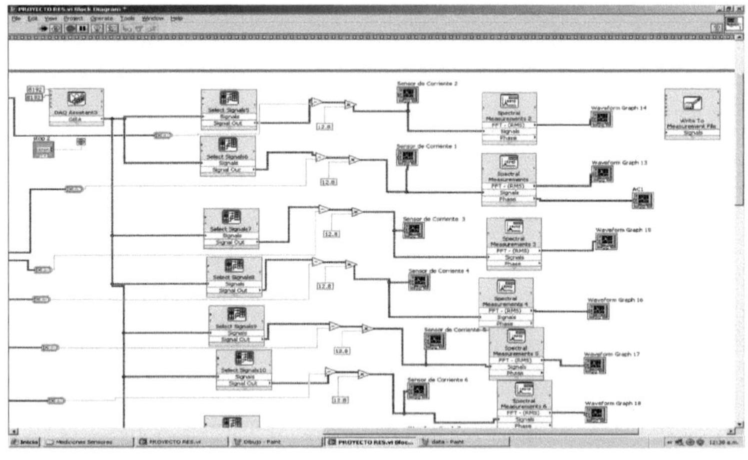

Figure 4.39 Image 2 of the program created in LabVIEW.

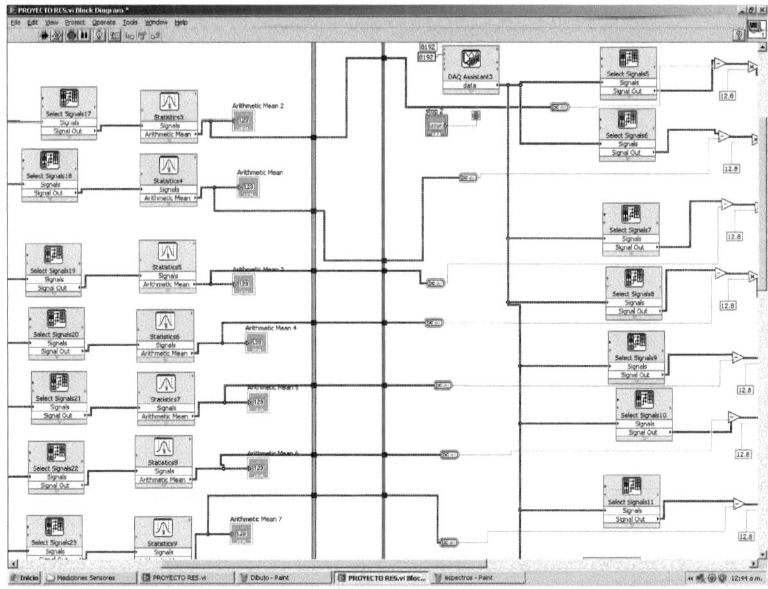

Figure 4.40 Image 3 of the program created in LabVIEW.

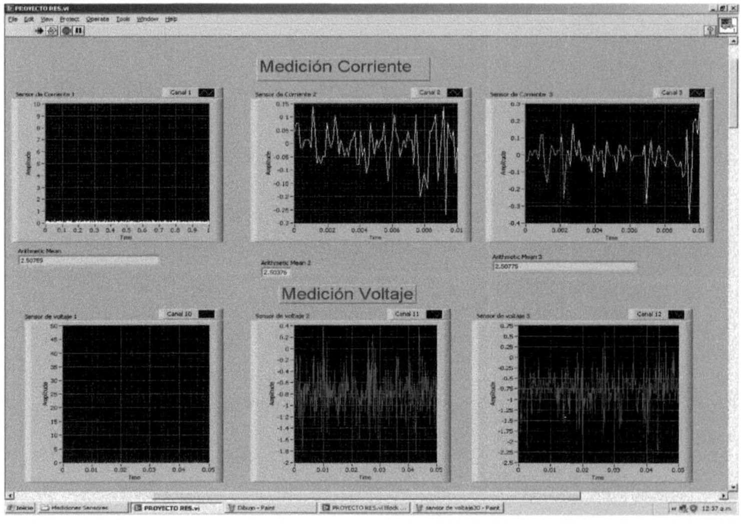

Figure 4.41 Front panel showing the graphs that will display the stator current and voltage measurements.

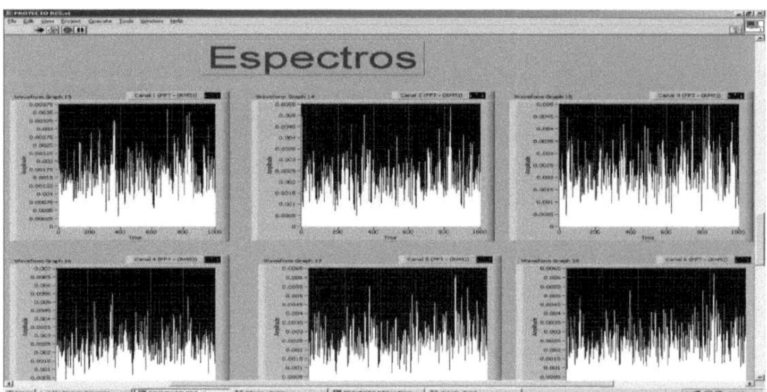

Figure 4.42 Front panel showing the graphs that will display the measured frequencies and magnitudes of the stator signals.

# CHAPTER 5.- RESULTS OBTAINED

In order to obtain results in the project, some tests are carried out and visualised graphically in LabVIEW. The tests consist of applying voltages, frequencies and different waveforms to the electrical machine and observing the behaviour in the LabVIEW graphs.

## 5.1 Test 1

For the first test, the power supply of the electrical machine was configured as follows:

Voltage: 5V AC

Frequency: 60 HZ

Source output phases: Three-phase

Output Signal Waveform: Pure sine wave

We then proceed to visualise the graphical visual aspects in LabVIEW such as Voltages, Currents, Frequencies and magnitudes.

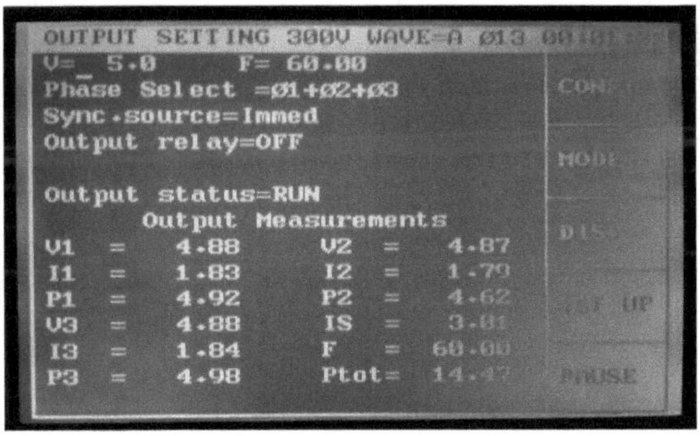

Figure 5.1 AC source programmed at 5 V, 60 Hz and 3 phases.

Graphical results of test 1

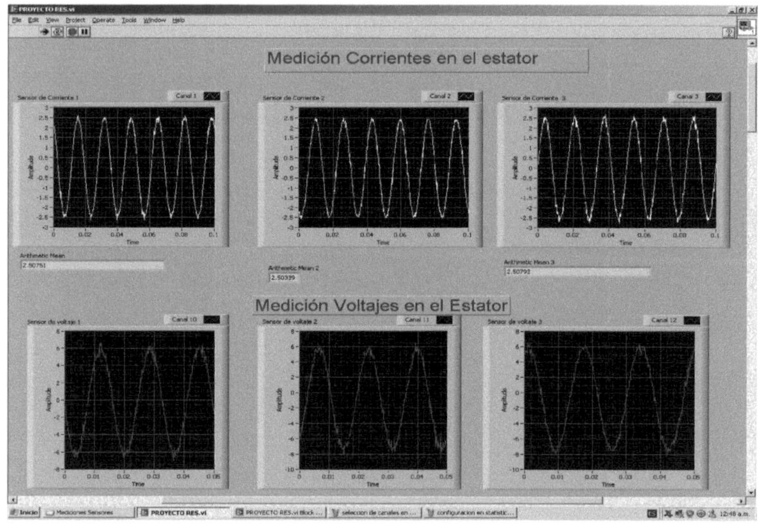

Figure 5.2 Graphical measurements of stator voltages and currents in test 1.

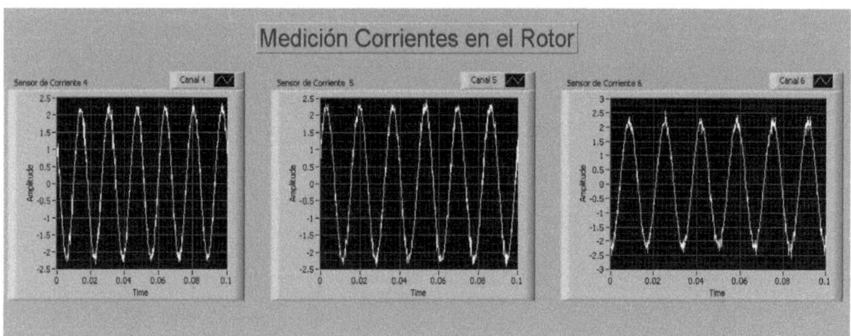

Figure 5.3 Graphical measurements of rotor currents in test 1.

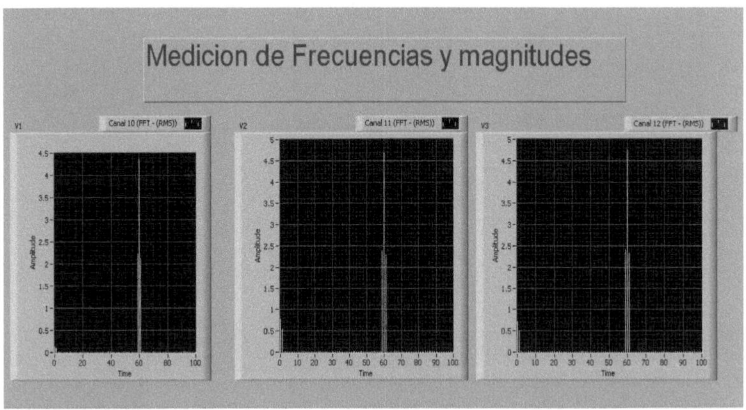


Figure 5.4 Frequency and magnitude measurements in stator voltage test 1

## 5.2 Test 2

For the second test, the power supply of the electrical machine was configured as follows:

Voltage: 5V AC

Frequency: 60 HZ

Source output phases: Three-phase

Output Signal Waveform: Sine wave with harmonics.

We then proceed to visualise the graphical visual aspects in LabVIEW such as Voltages, Currents, Frequencies and magnitudes.

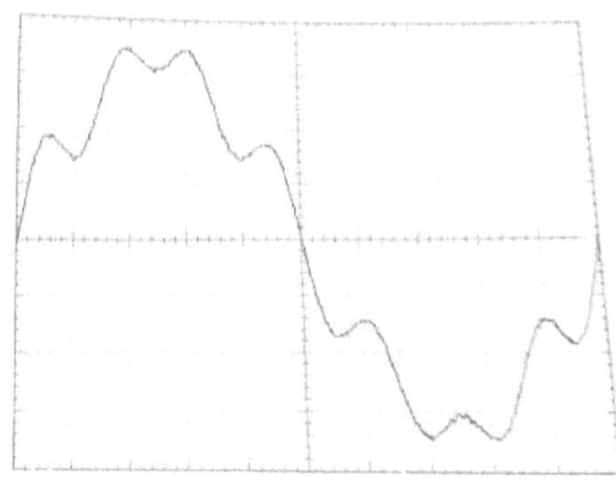



| Harmonic | % |
|---|---|
| 2 | 2.3 |
| 5 | 9.8 |
| 7 | 15.8 |
| 8 | 2.5 |

Figure 5.5 Table with the number of harmonics and their percentage level and waveform applied for test 2.

Graphical results of test 2

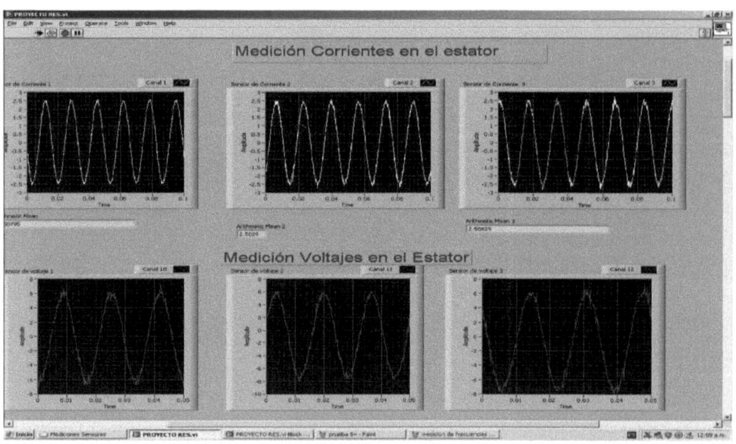



Figure 5.6 Stator voltage and current measurements in test 2.

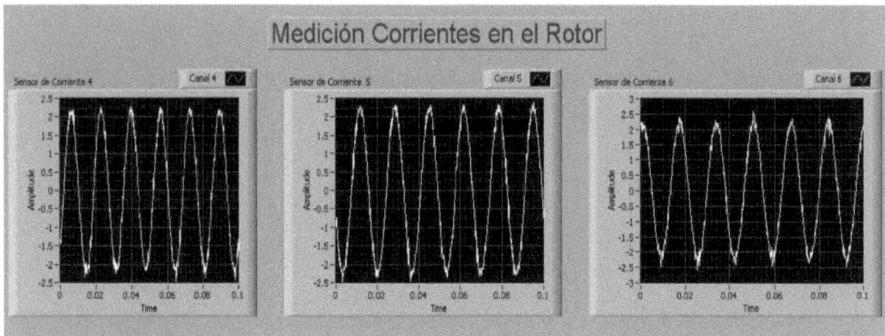


Figure 5.7 Graphical measurements of rotor currents in test 2.

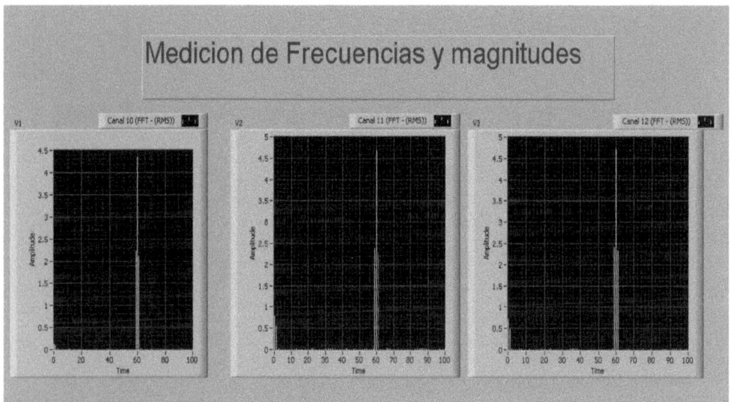


Figure 5.8 Frequency and magnitude measurements in stator voltage test 2.

## 5.3 Test 3

For the third test, the power supply of the electrical machine was configured as follows:

Voltage: 7V AC

Frequency: 300 HZ

Source output phases: Three-phase

Output Signal Waveform: Pure sine wave.

We then proceed to visualise the graphical visual aspects in LabVIEW such as Voltages, Currents, Frequencies and magnitudes.

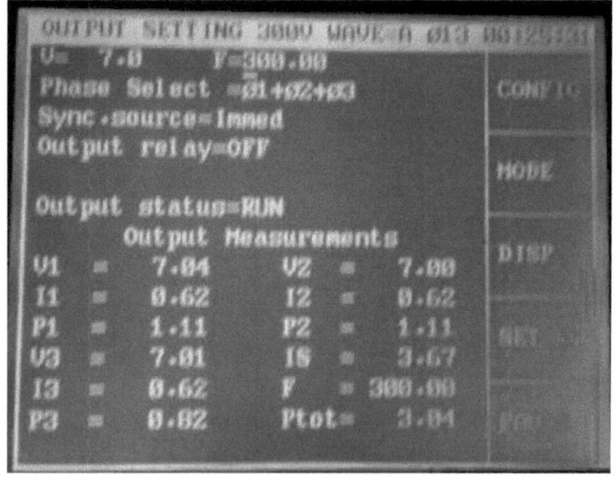

Figure 5.9 AC source programmed at 7 V, 300 Hz and 3 phases.

Results Test 3 graphs

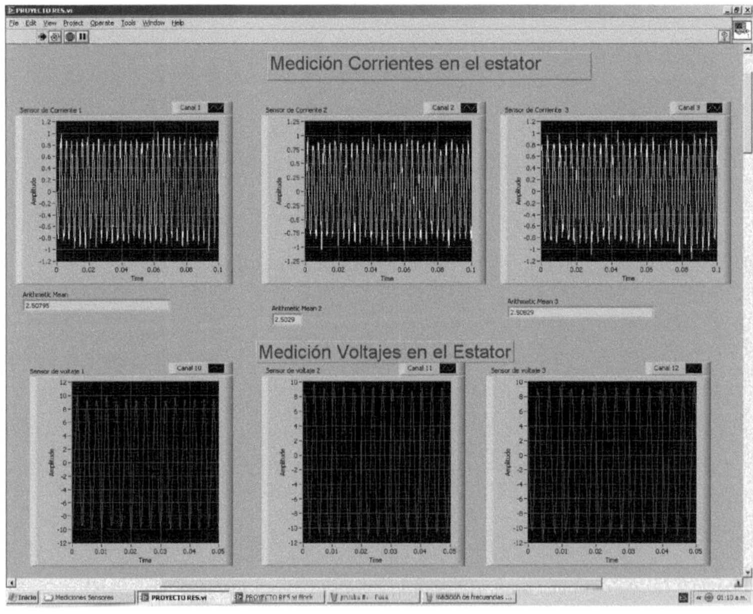

Figure 5.10 Graphical measurements of stator voltages and currents in test 3.

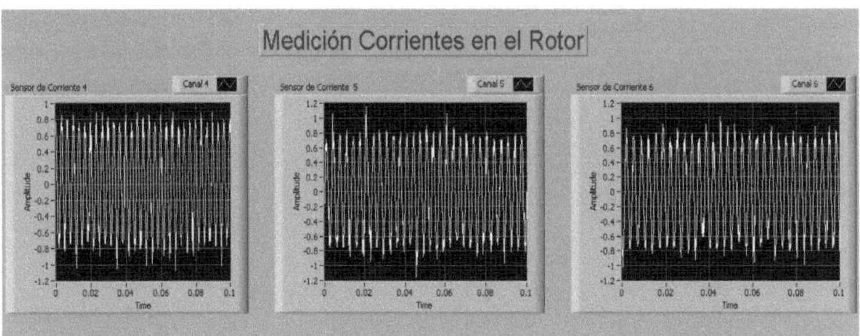

Figure 5.11 Graphical measurements of rotor currents in test 3.

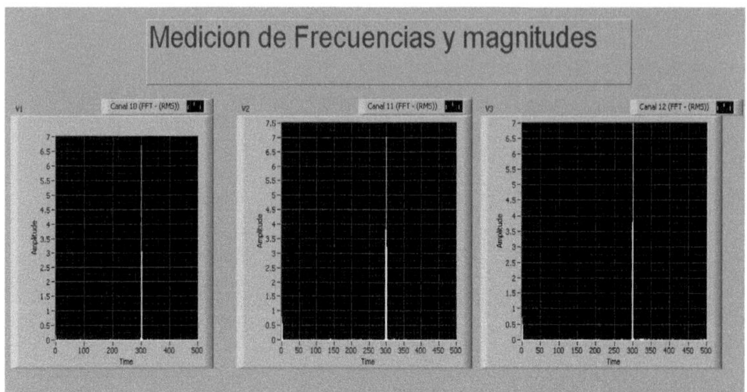


Figure 5.12 Frequency and magnitude measurements in stator voltage test 3.

**Frequency sweep test**

In the frequency sweep test, different voltages (10, 15 and 20 V) at different frequencies (from 45 Hz to 1000 Hz) were applied to the electrical machine.

Then we proceeded to visualize the visual aspects in LabVIEW graphics as voltages, currents, frequencies, magnitudes of the angles, once the tests were performed and obtained a database were made different graphs such as: Voltage-frequency graphs in the stator and rotor , angle-frequency graphs of voltage in the stator and rotor, stator current-frequency graphs, rotor current-frequency graphs and their respective angles, impedance and admittance graphs and in the stator and rotor and their respective angles and graphs of the transformation ratio of stator current between rotor current and their respective angles. The graphs are shown below for analysis.

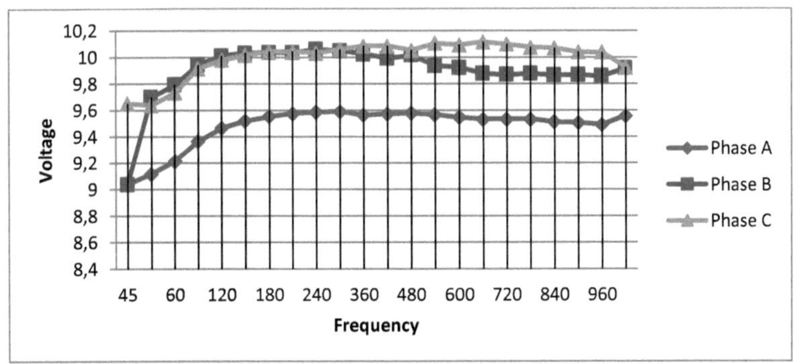

Figure 5.13 Stator voltage-frequency plot at 10 v.

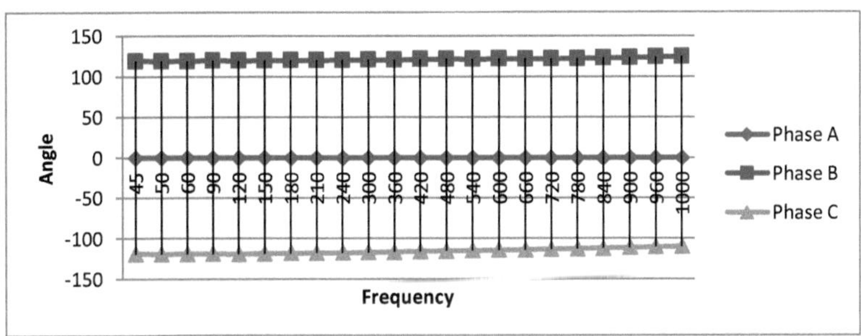

Figure 5.14 Angle-frequency plot of stator voltage at 10 v.

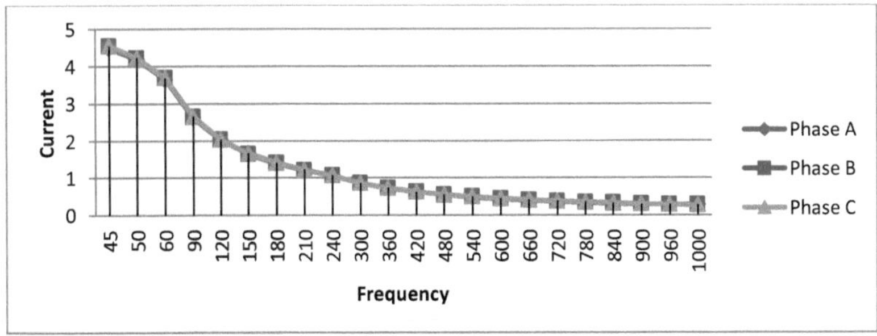

Figure 5.15 Stator current-frequency graph at 10V.

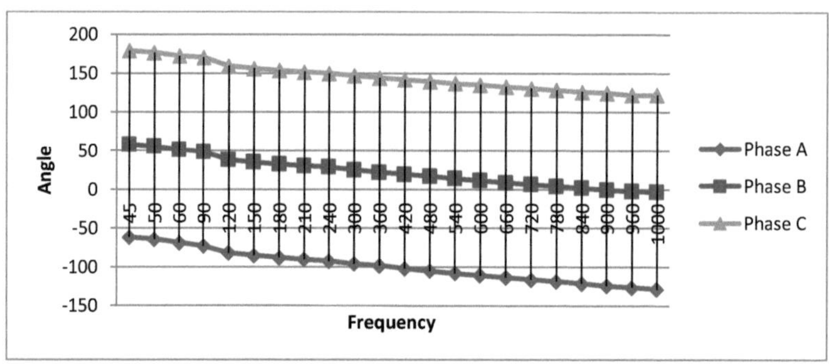

Figure 5.16 Angle-frequency plot of stator current at 10 v.

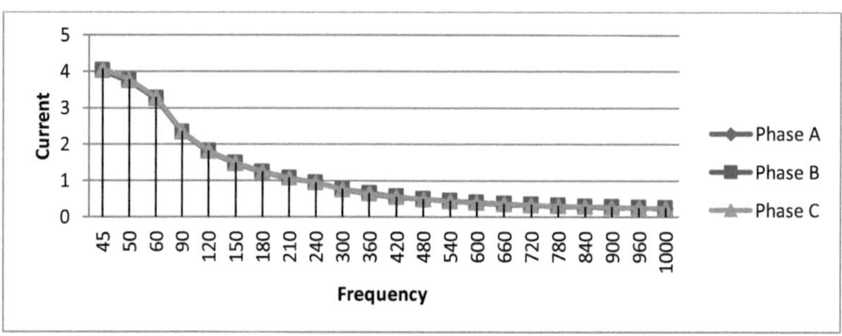

Figure 5.17 Rotor current-frequency graph at 10V.

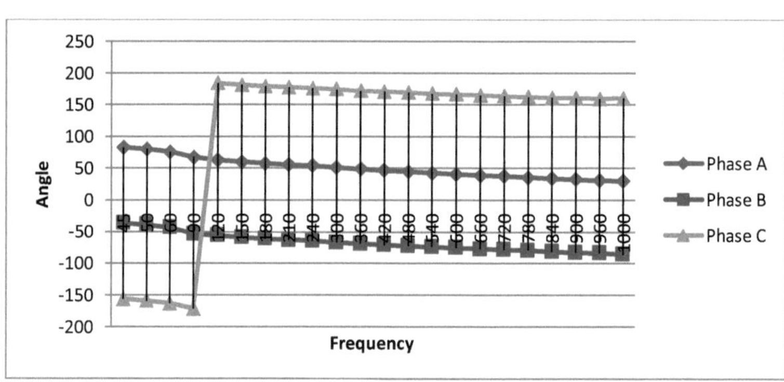

Figure 5.18 Rotor current angle-frequency graph at 10 v.

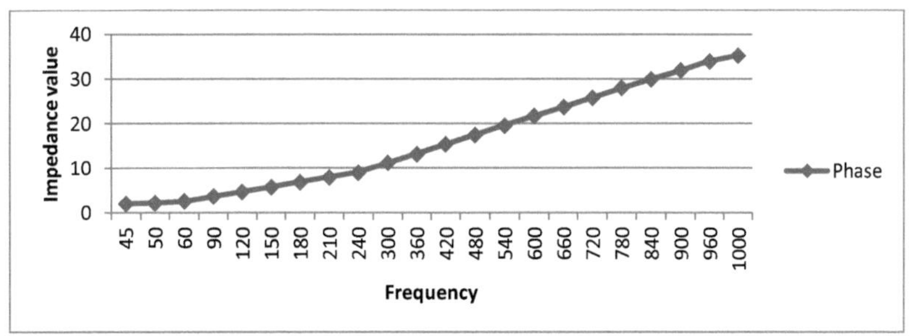

Figure 5.19 Stator impedance graph Ve/Ie 10 v.

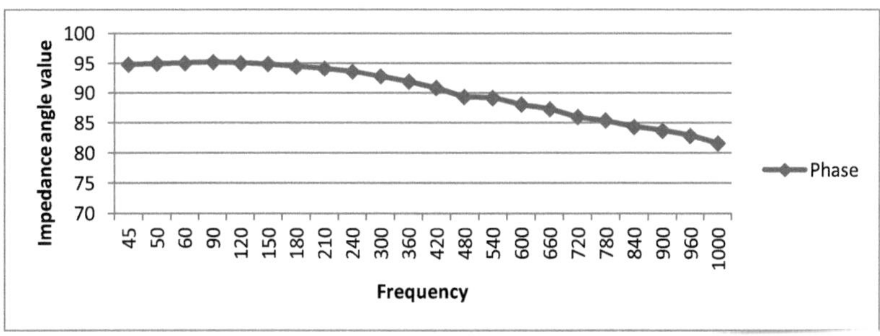

Figure 5.20 Plot of stator impedance angles Ve/Ie 10 v.

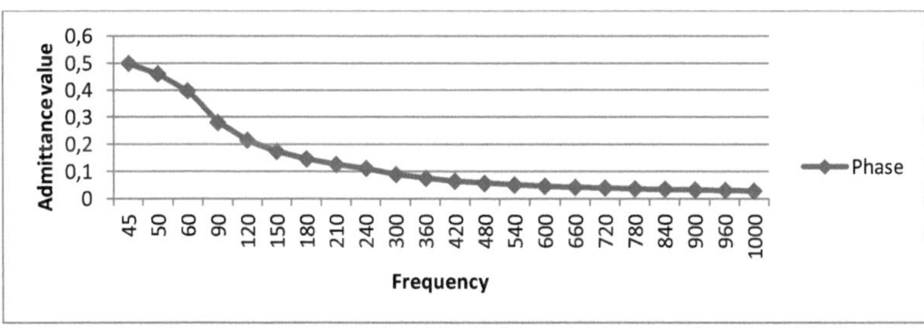

Figure 5.21 Stator admittance graph Ve/Ie 10 v.

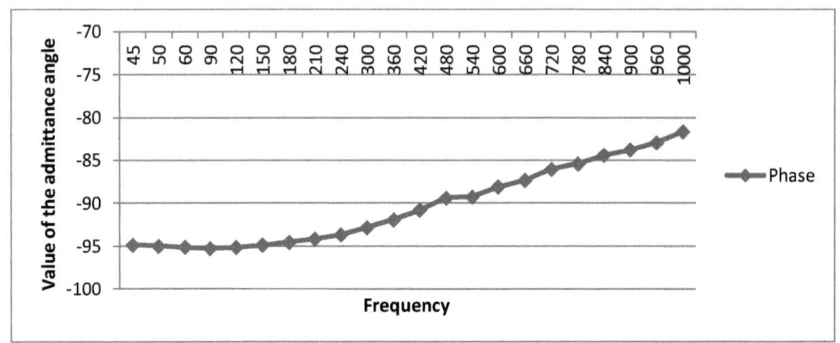

Figure 5.22 Plot of stator admittance angles Ve/Ie 10 v.

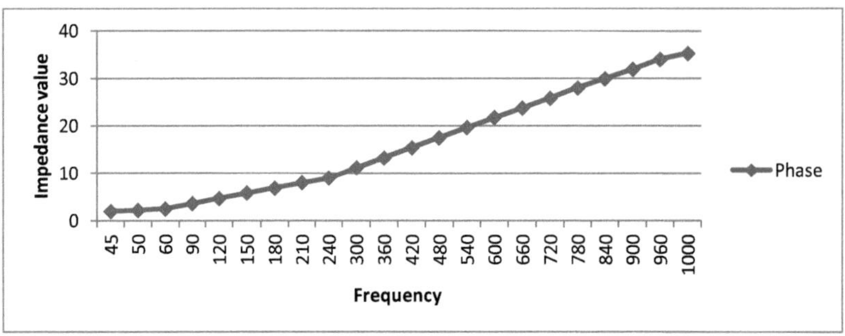

Figure 5.23 Rotor impedance graph Ve/Ir 10 V.

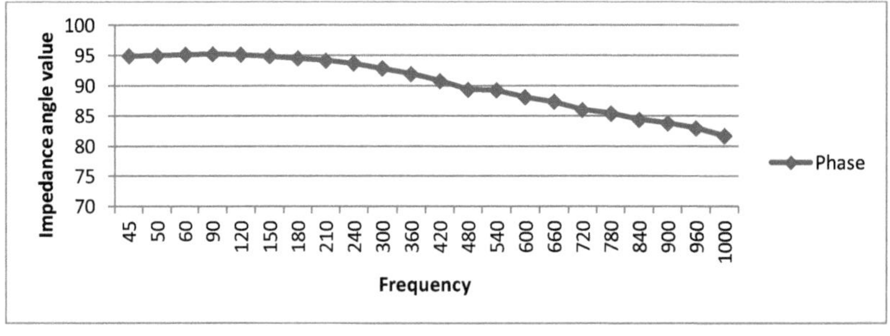

Figure 5.24 Plot of impedance angles at rotor Ve/Ir 10 V.

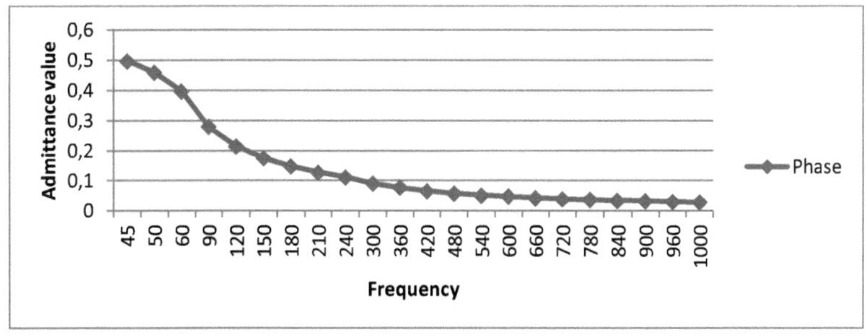

Figure 5.25 Rotor admittance graph Ve/Ir 10 V

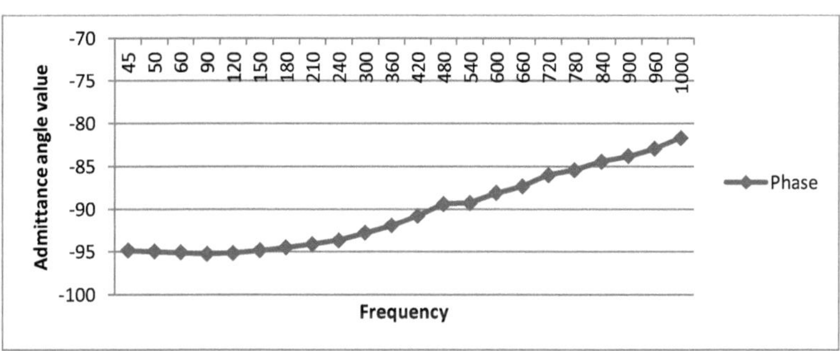

Figure 5.26 Plot of rotor admittance angles Ve/Ir at 10V.

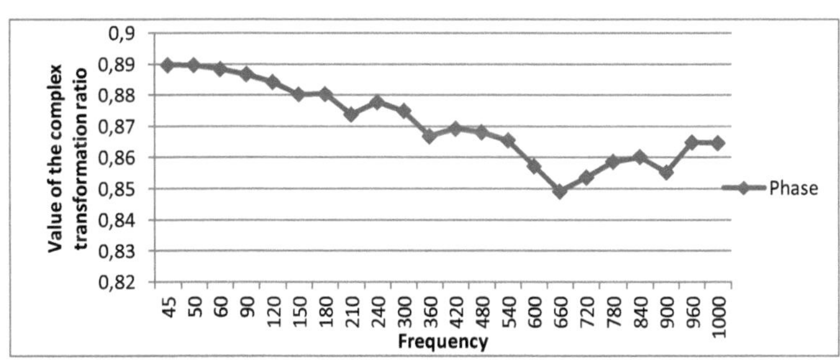

Figure 5.27 Graph of Ir/Ie complex transformation ratio at 10 V.

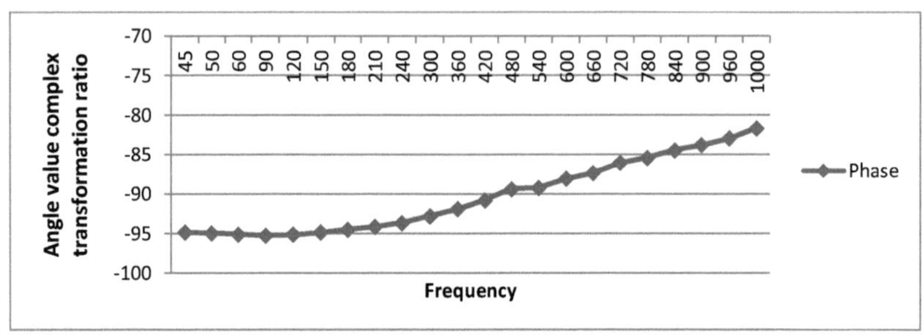

Figure 5.28 Plot of the angles of the complex transformation ratio Ir/Ie at different frequencies at 10V.

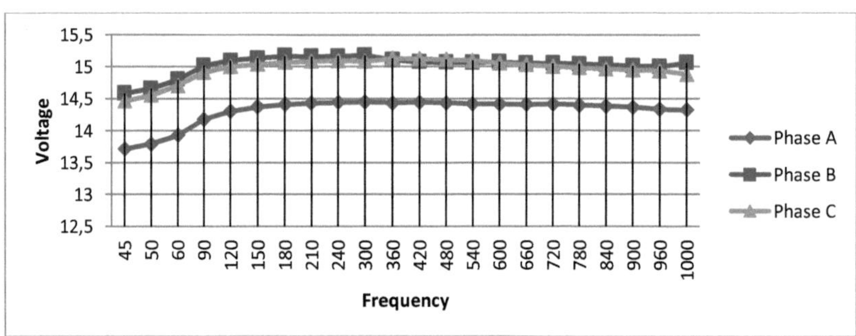

Figure 5.29 Stator voltage-frequency graph at 15 v.

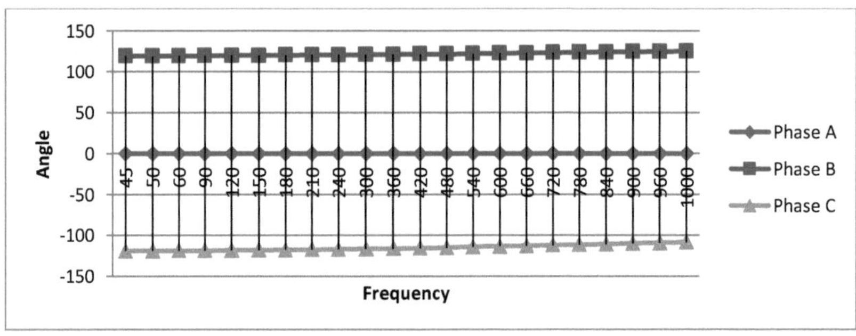

Figure 5.30 Angle-frequency plot of stator voltage at 15 v.

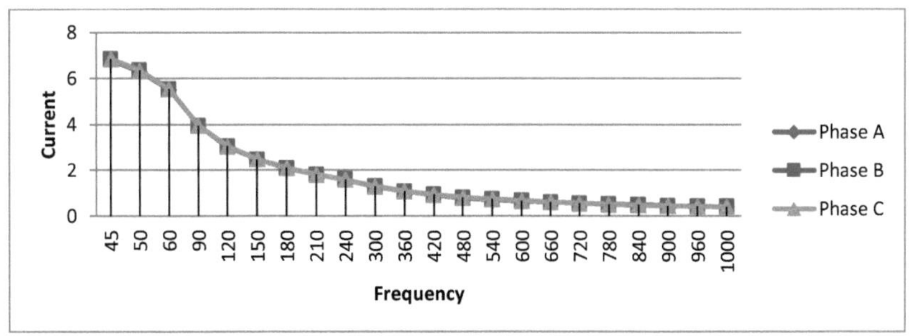

Figure 5.31 Stator current-frequency graph at 15V.

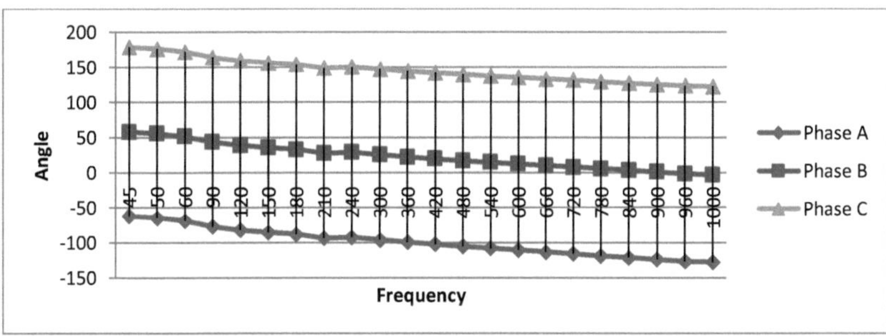

Figure 5.32 Angle-frequency graph of stator current at 15 v.

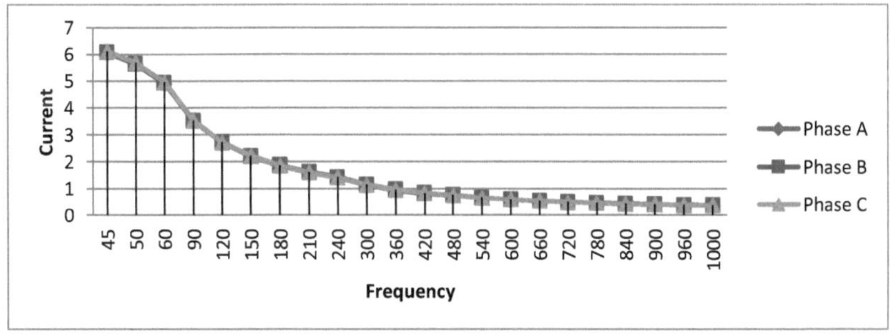

Figure 5.33 Rotor current-frequency graph at 15V.

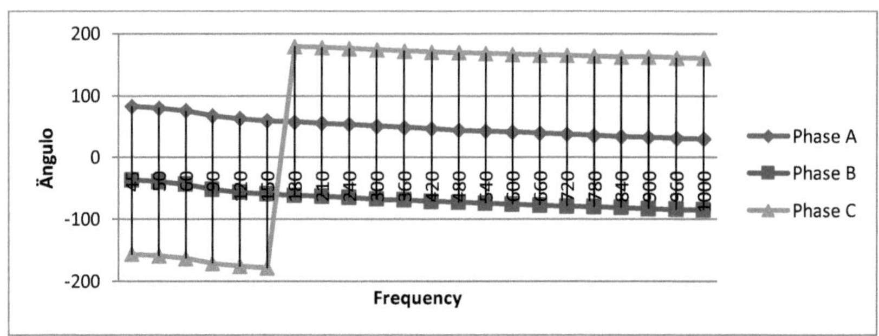

Figure 5.34 Rotor current angle-frequency graph at 15V.

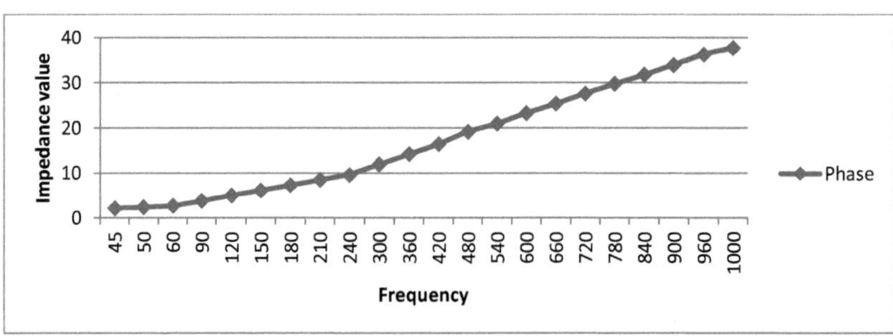

Figure 5.35 Stator impedance graph Ve/Ie at 15V.

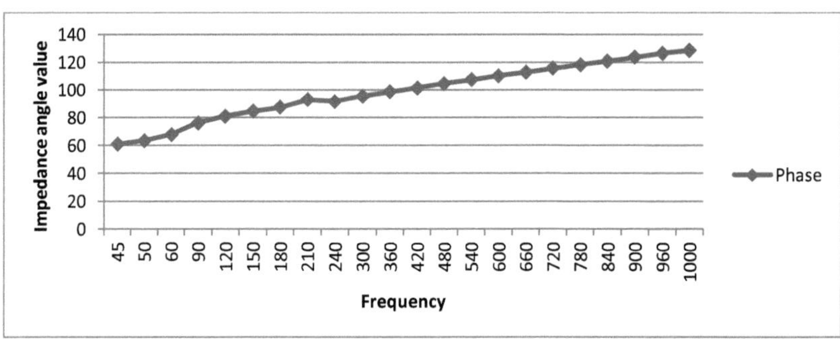

Figure 5.36 Plot of stator impedance angles Ve/Ie at 15V.

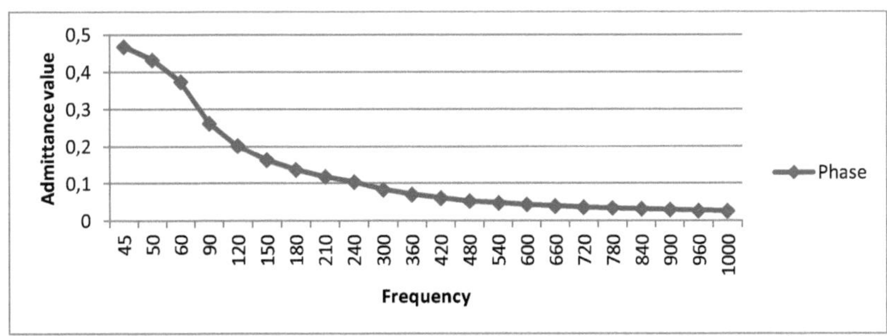

Figure 5.37 Stator admittance graph Ve/Ie at 15V.

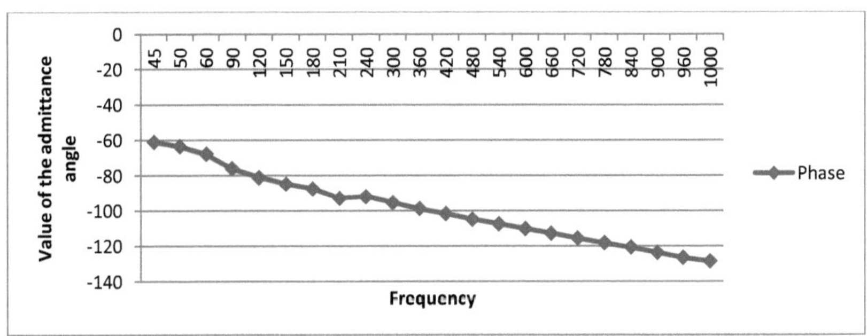

Figure 5.38 Plot of stator admittance angles Ve/Ie 15V.

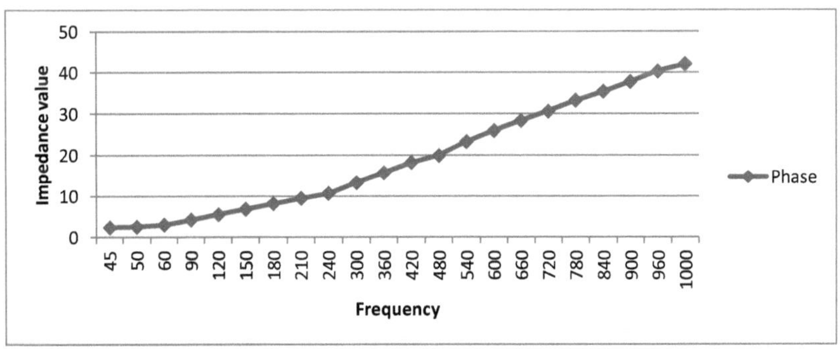

Figure 5.39 Impedance plot of rotor impedance Ve/Ir at 15V.

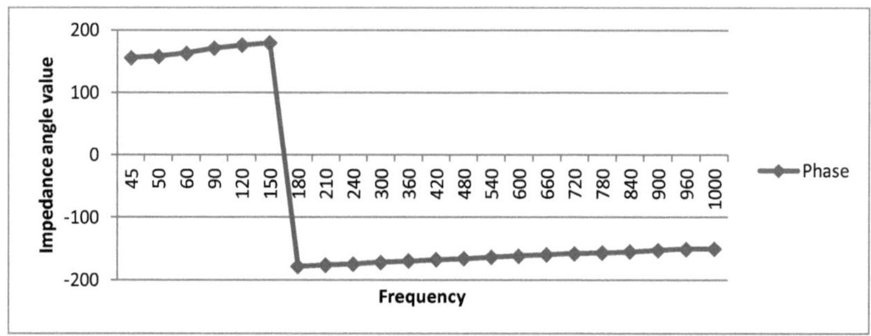

Figure 5.40 Plot of the impedance angles in the Ve/Ir rotor at 15 V.

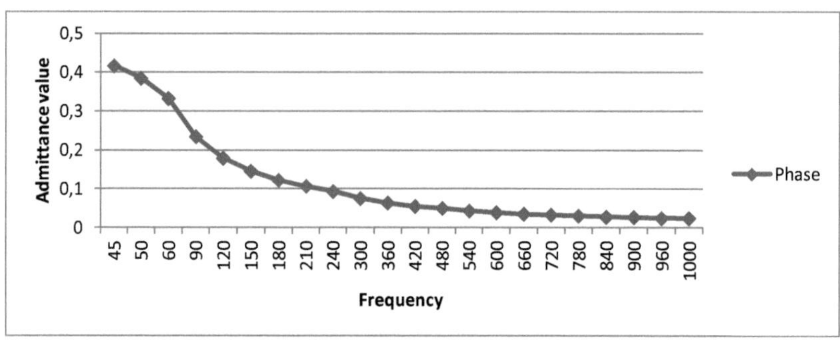

Figure 5.41 Plot of rotor admittance Ve/Ir at 15V.

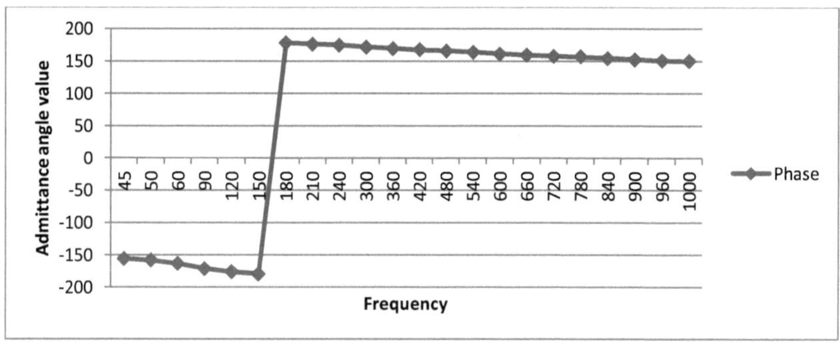

Figure 5.42 Plot of rotor admittance angles Ve/Ir at 15V.

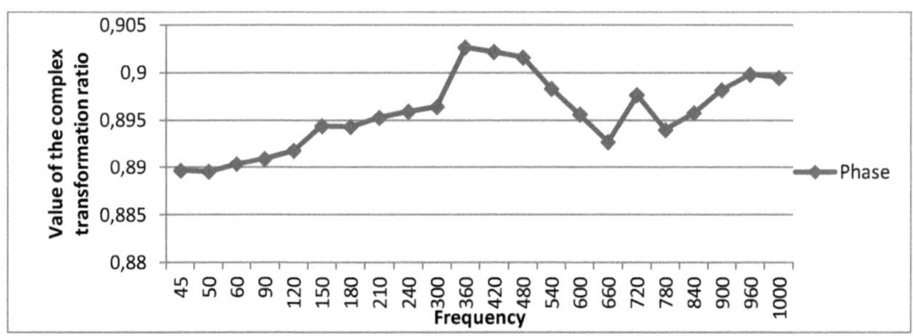

Figure 5.43 Plot of the Ir/Ie complex transformation ratio at 15 V.

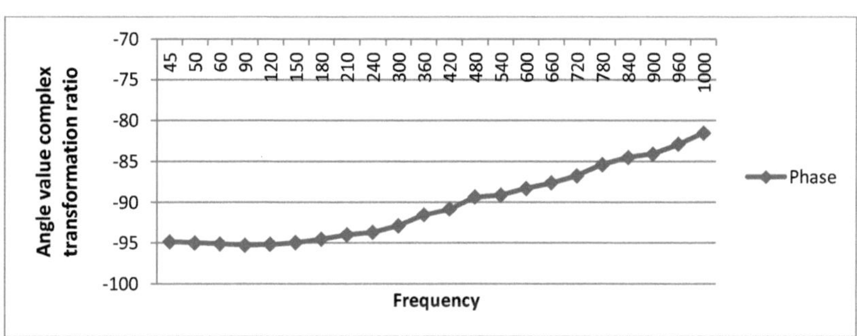

Figure 5.44 Angle plot of the complex transformation ratio Ir/Ie at 15 V.

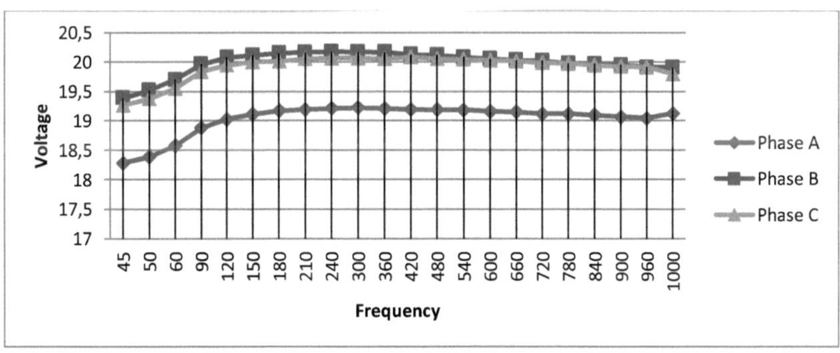

Figure 5.45 Stator voltage-frequency plot at 20 v.

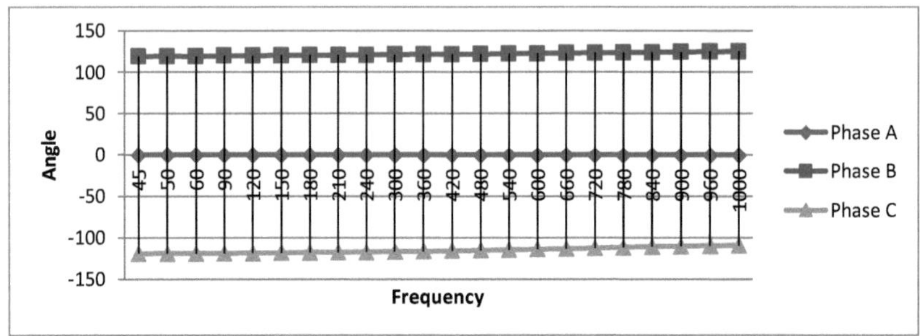

Figure 5.46 Angle-frequency plot of stator voltage at 20 v.

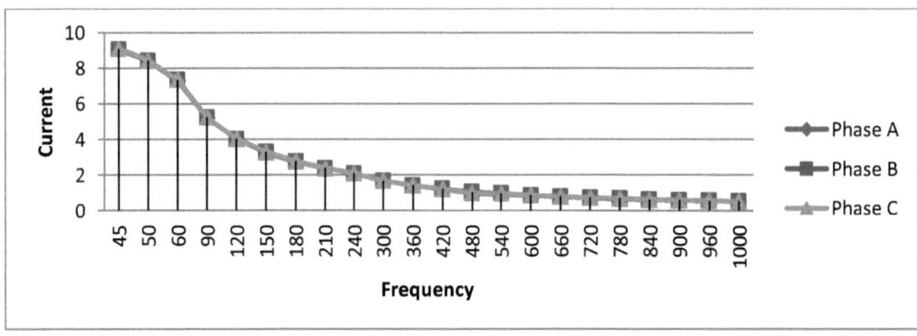

Figure 5.47 Stator current-frequency graph at 20V.

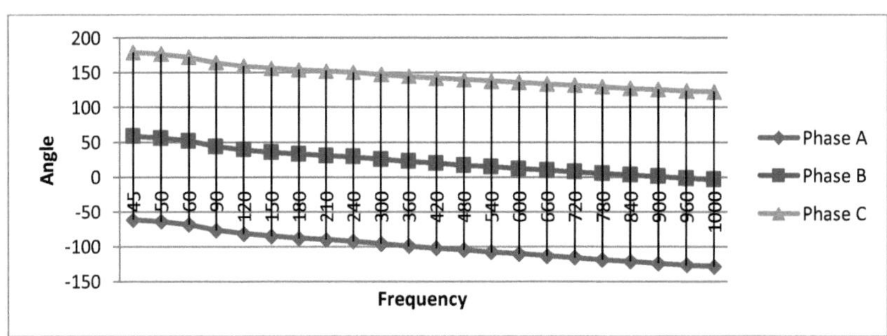

Figure 5.48 Angle-frequency plot of stator current at 20 v.

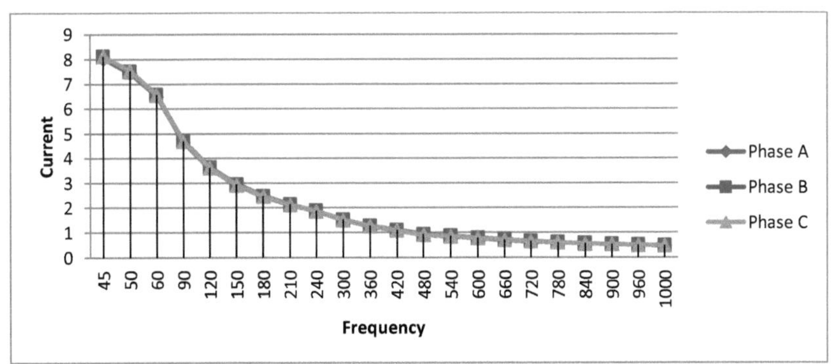

Figure 5.49 Rotor current-frequency graph at 20V.

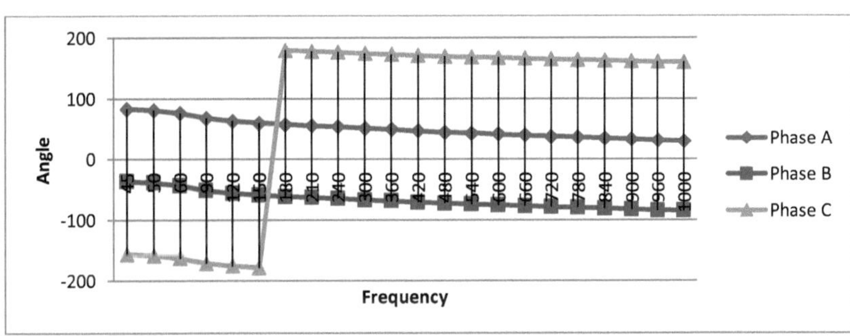

Figure 5.50 Rotor current angle-frequency graph at 20V.

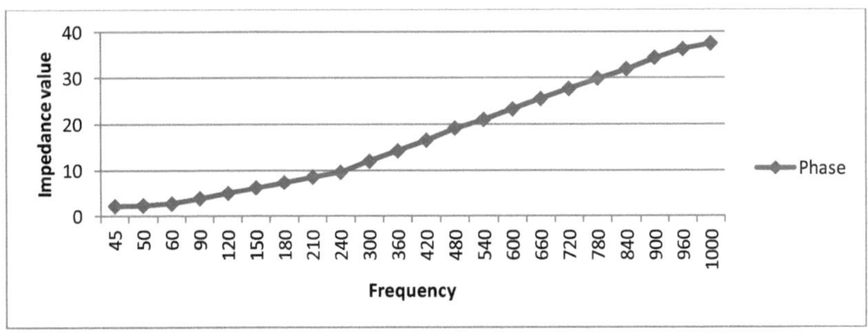

Figure 5.51 Stator impedance graph Ve/Ie at 20V.

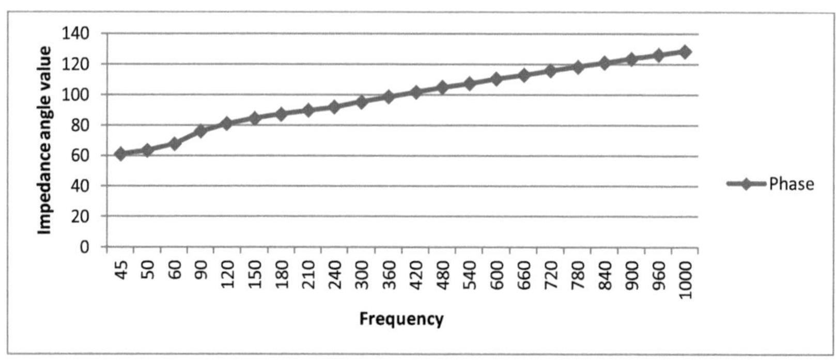

Figure 5.52 Plot of stator impedance angles Ve/Ie at 20V.

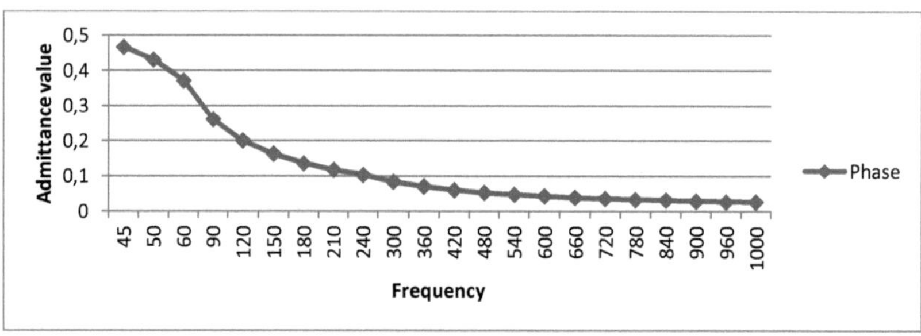

Figure 5.53 Stator admittance graph Ve/Ie at 20V.

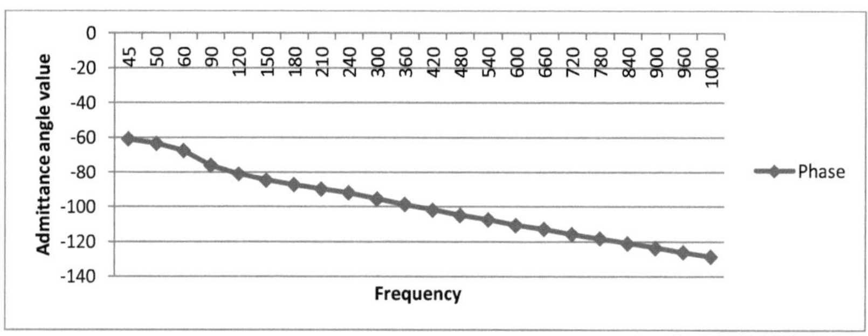

Figure 5.54 Plot of stator admittance angles Ve/Ie at 20V.

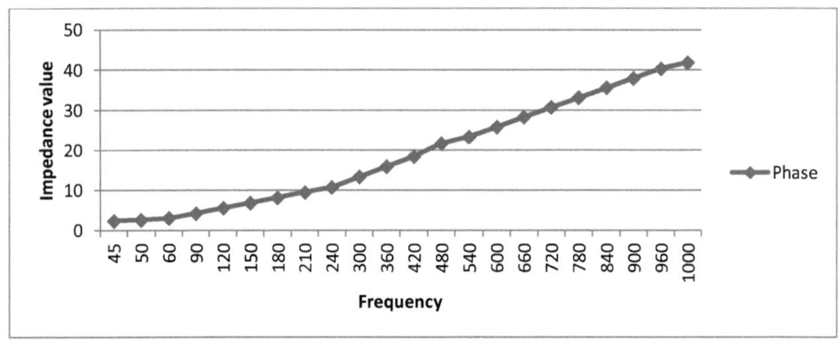

Figure 5.55 20 V Ve/Ir rotor impedance plot.

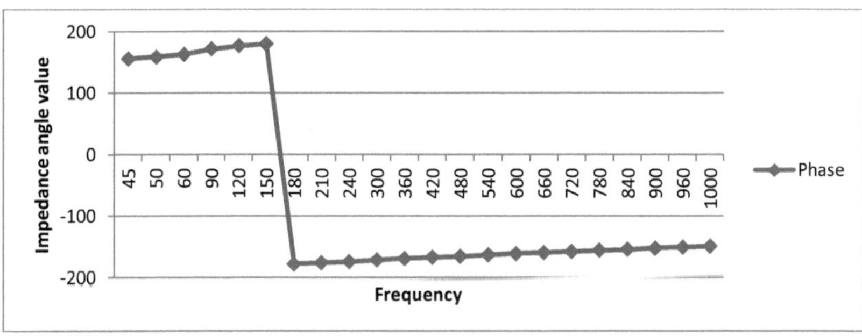

Figure 5.56 Plot of the impedance angles in the Ve/Ir 20 V rotor.

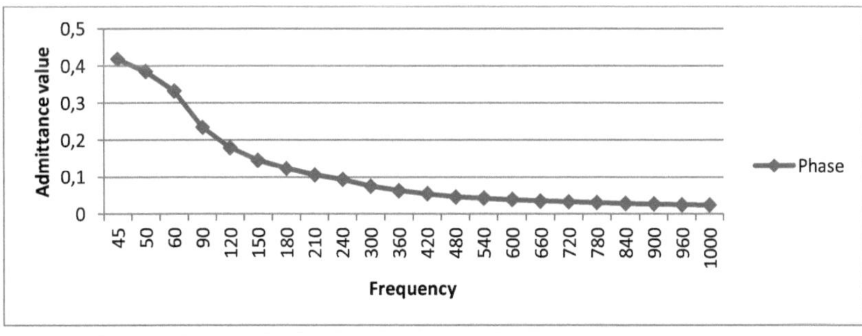

Figure 5.57 20 V Ve/Ir rotor admittance graph.

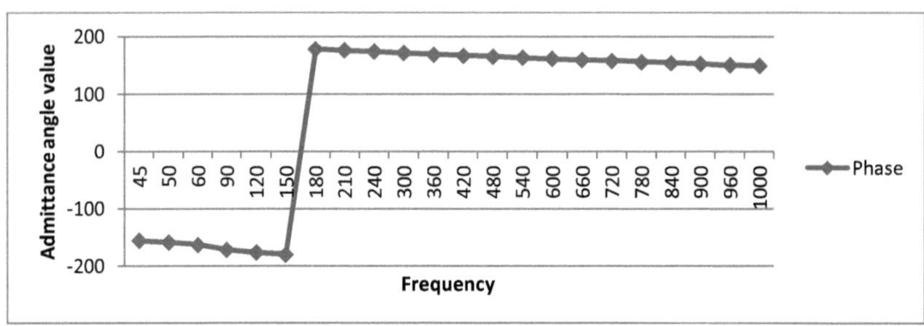

Figure 5.58 Plot of the admittance angles of the Ve/Ir 20 V rotor.

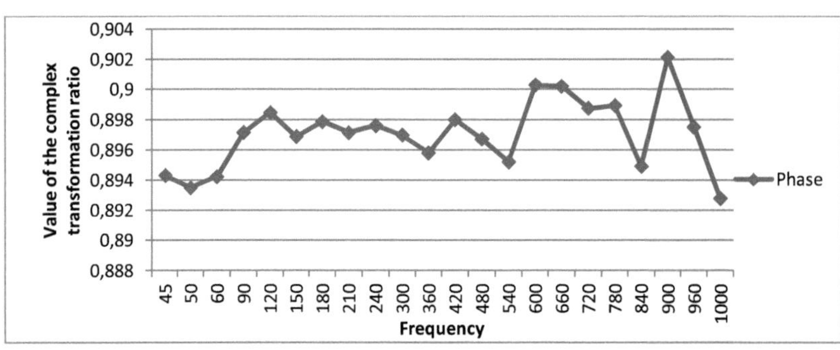

Figure 5.59 Plot of the Ir/Ie complex transformation ratio at 20 V.

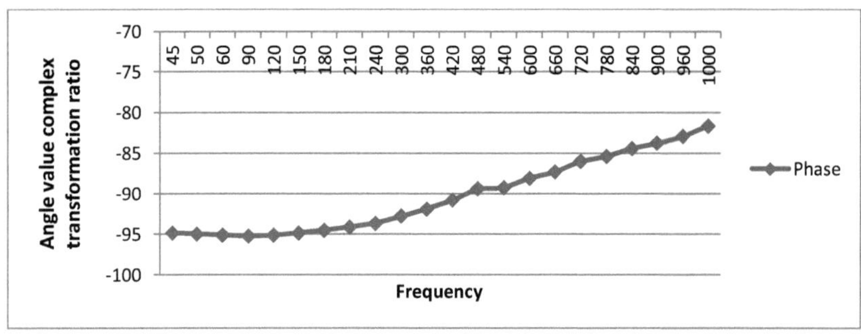

Figure 5.60 Angle plot of the Ir/Ie complex transformation ratio at 20 V

**Harmonic contamination test**

In the harmonic contamination test, distorted waveforms with harmonics were applied to the electrical machine to see the behaviour of the variables of the electrical machine.

The waveforms that were chosen were those with the highest percentage of harmonic contamination provided by the *Chroma* Model 6590 programmable AC source.

The distorted waveform and the distortion values of the harmonics are shown below.

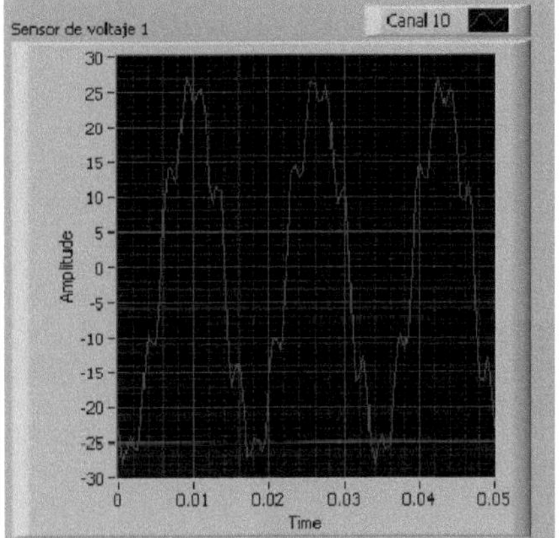


| Harmonics | % |
|---|---|
| 2 | 2.3 |
| 5 | 9.8 |
| 7 | 15.8 |
| 8 | 2.5 |

Figure 5.61 DST014 distorted waveform and table of harmonics and their percentage provided by the source.

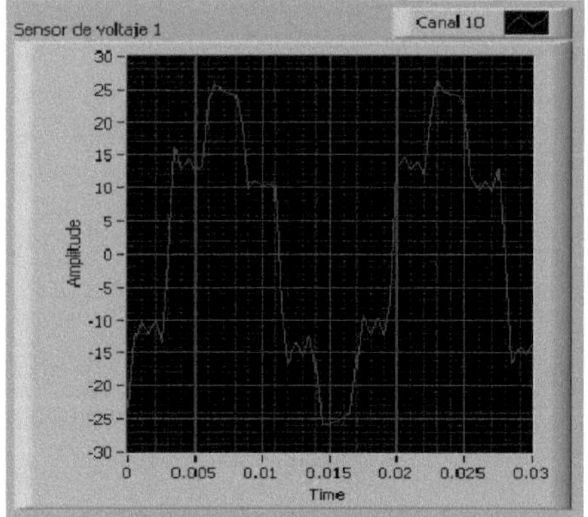


| Harmonics | % |
|---|---|
| 5 | 2.3 |
| 7 | 9.8 |
| 11 | 15.8 |
| 13 | 2.5 |

Figure 5.62 DST030 distorted waveform and table of harmonics and their percentage provided by the source.

To verify that the source was delivering the generated harmonics to the output, a FLUKE 435 series II power quality meter was used, this device is capable of analysing harmonics and their distortion percentages.

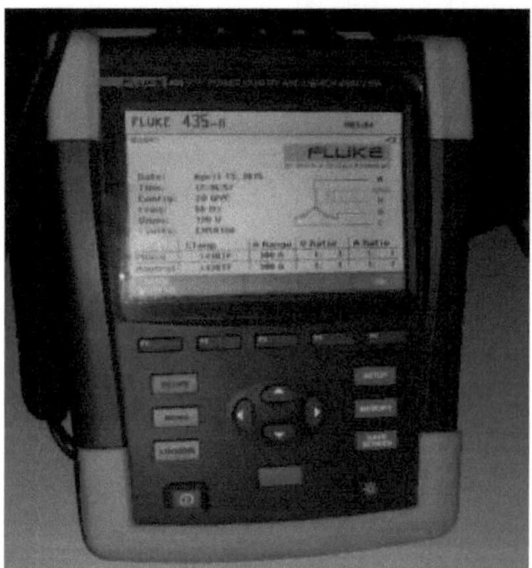

Figure 5.63 Power quality meter used for verification of induced harmonics.

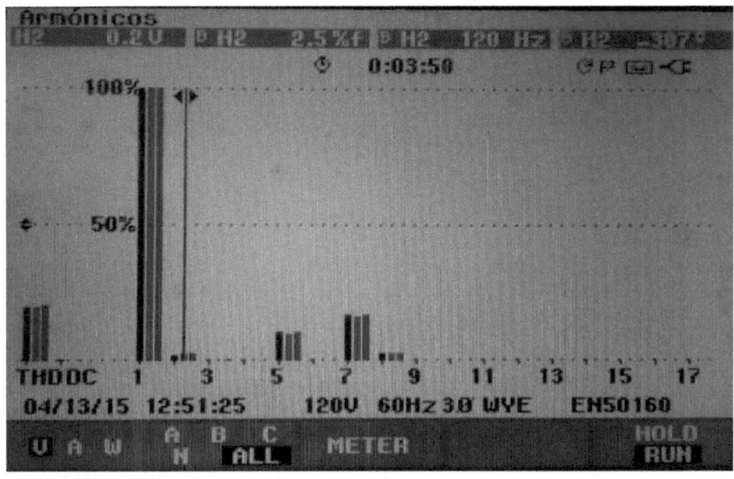

Figure 5.64 Harmonics in the DST014 waveform sampled on the FLUKE 435
series II power quality meter.

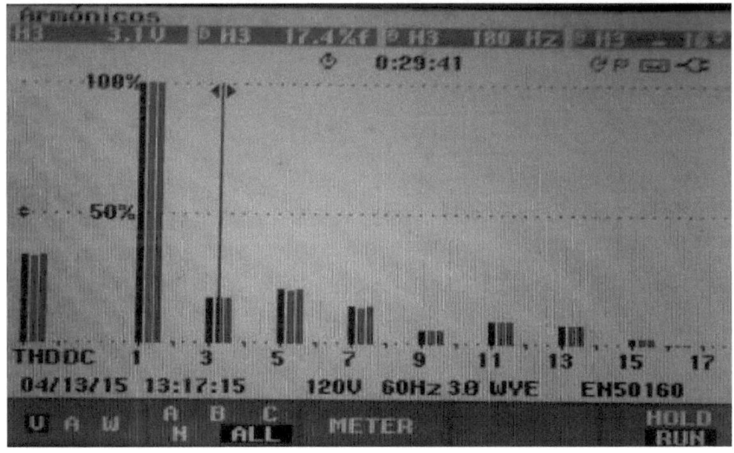

Figure 5.65 Harmonics in the DST030 waveform sampled on the FLUKE 435
series II power quality meter.

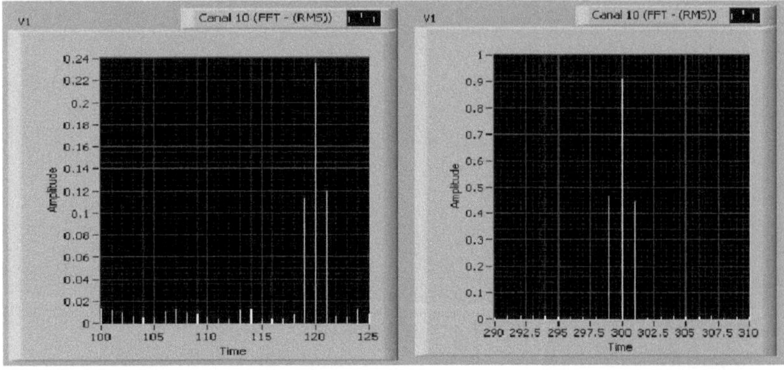

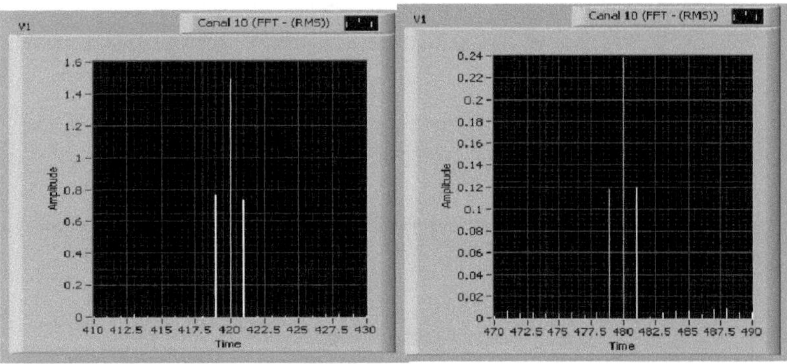

Figure 5.66 LabVIEW sampling of harmonics 2, 5, 7 and 8 of the distorted waveform DST014.

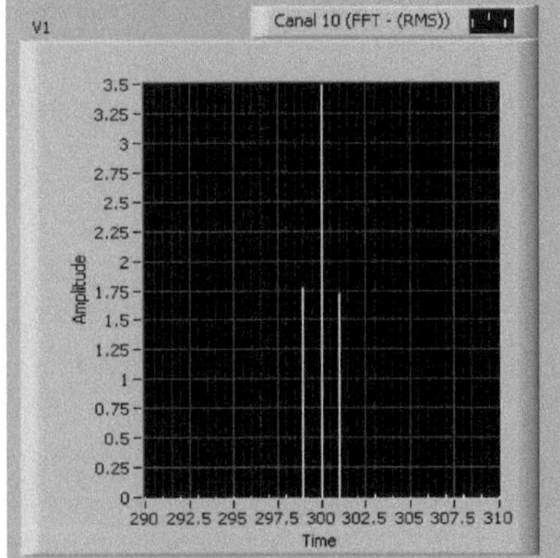 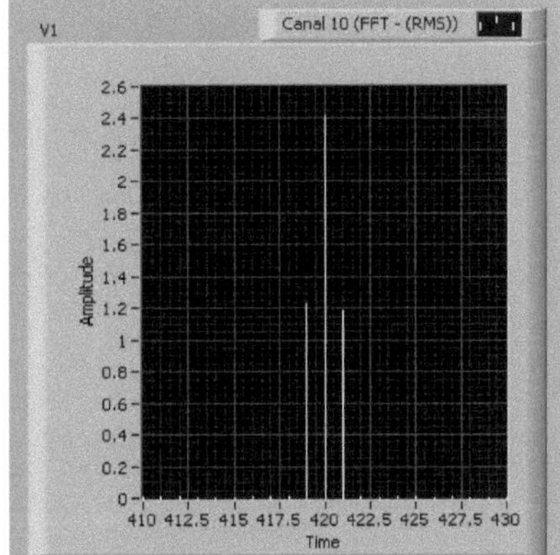

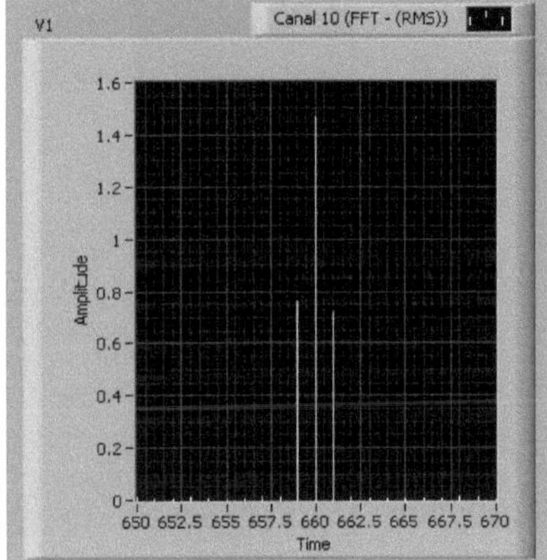 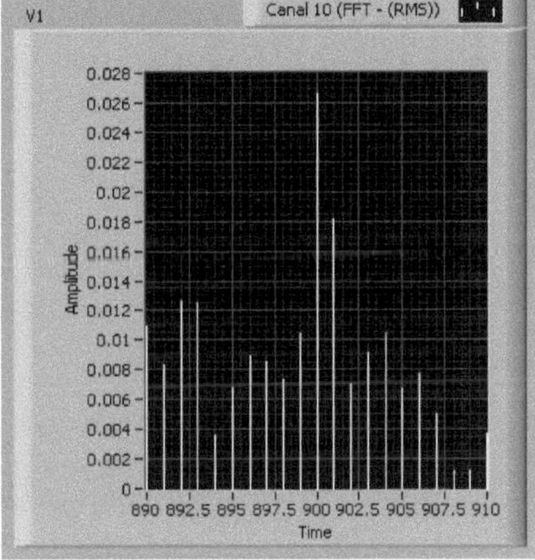

Figure 5.67 LabVIEW sampling of harmonics 5, 7, 11 and 13 of the distorted waveform DST030.

Once it was verified that the source correctly provides the harmonics and their percentage level, the tests previously performed were carried out. In the same way, a database was acquired in LabVIEW and the impedances and admittances in the electrical machine were plotted, as well as the complex transformation ratio Ir/Ie.

The results of the graphs are shown below.

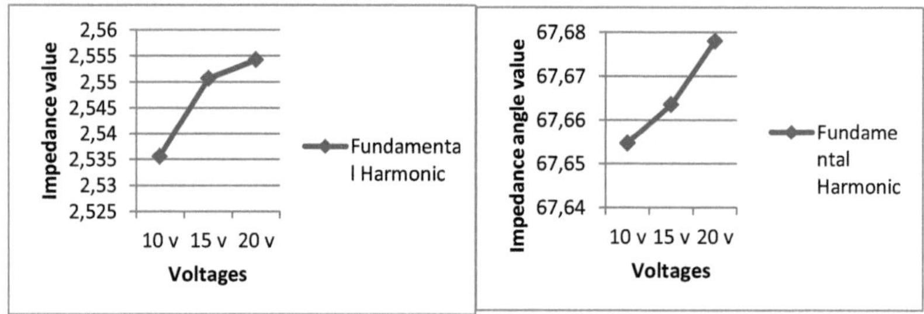

Figure 5.68 Plot of the stator impedance at the 60 Hz fundamental harmonic and the value of the angles with waveform DST 014.

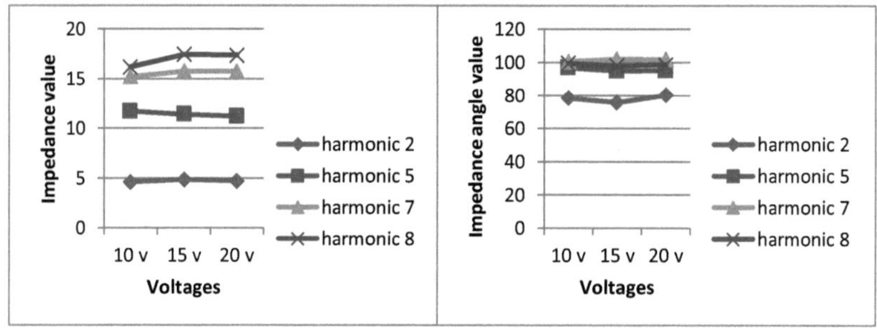

Figure 5.69 Plot of the stator impedance of the harmonics and the value of the angles with waveform DST014.

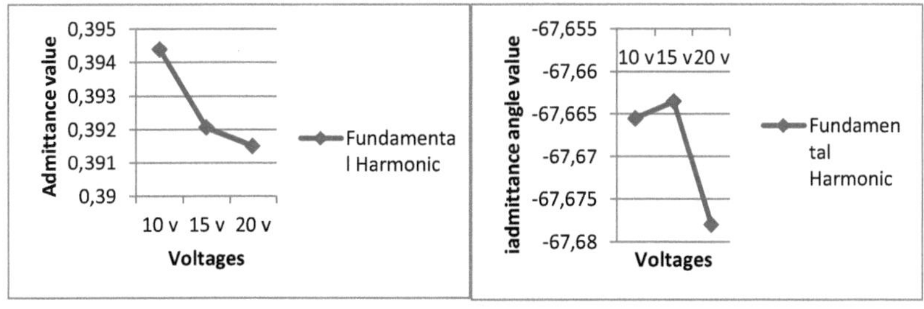

Figure 5.70 Plot of stator admittance at the 60 Hz fundamental harmonic and the value of the angles with waveform DST 014.

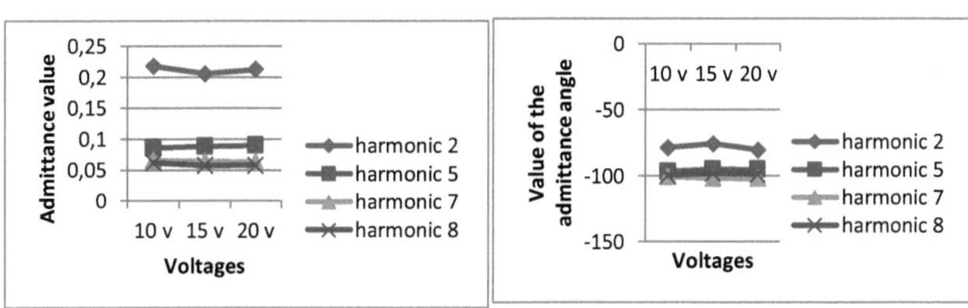

Figure 5.71 Plot of stator admittance at harmonics and angle value with waveform DST014.

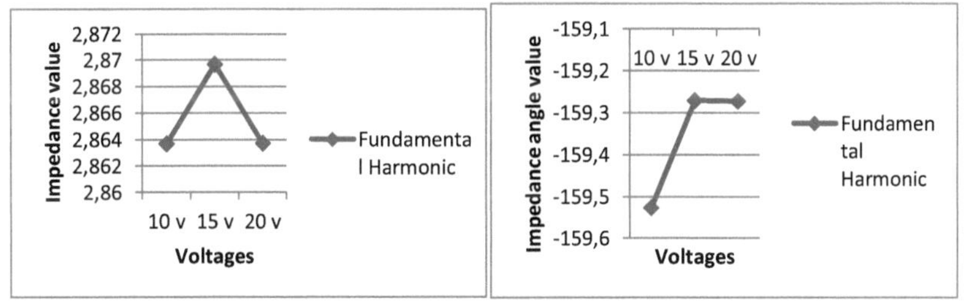

Figure 5.72 Plot of rotor impedance at the fundamental 60 Hz harmonic and the value of the angles with waveform DST 014.

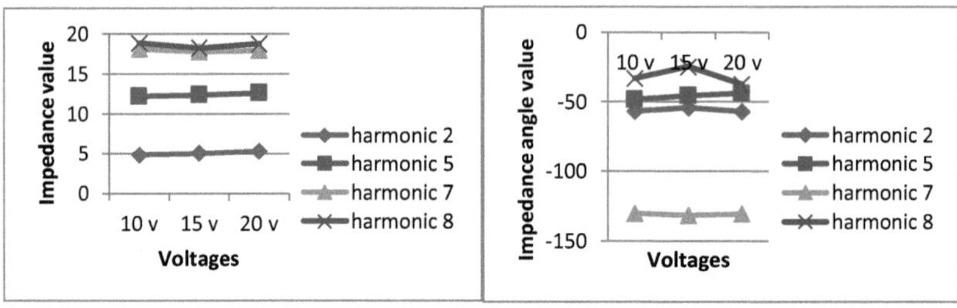

Figure 5.73 Plot of harmonic rotor impedance and angle value with waveform DST014.

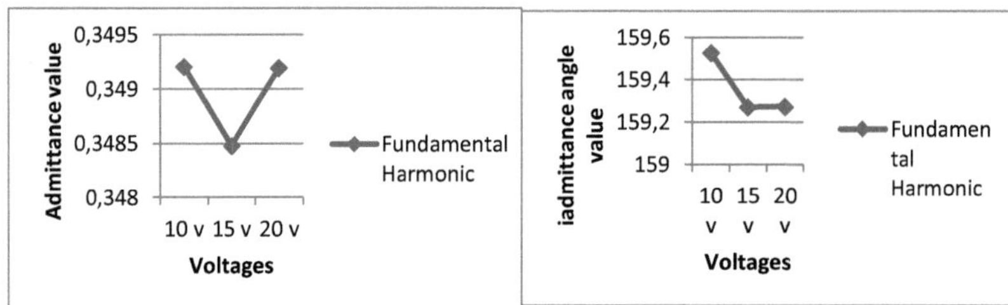

Figure 5.74 Plot of the rotor admittance at the fundamental 60 Hz harmonic and the value of the angles with waveform DST 014.

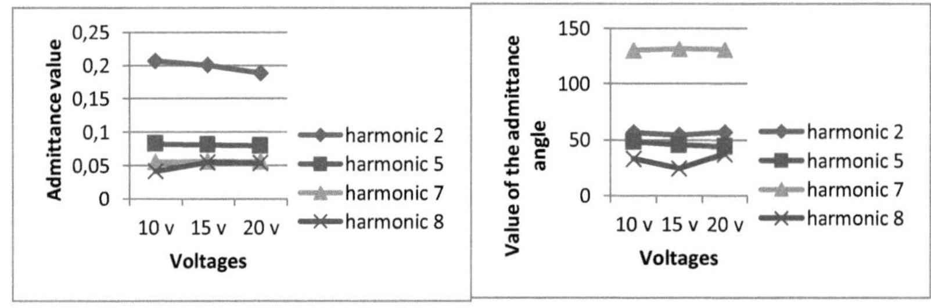

Figure 5.75 Plot of harmonic rotor admittance and angle value with waveform DST014.

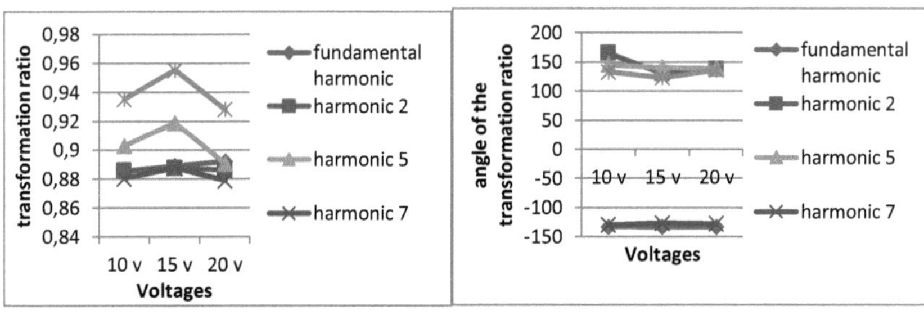

Figure 5.76 Plot of Ir/Ie complex transformer ratio with DST014 waveform and harmonic inductance at different voltages.

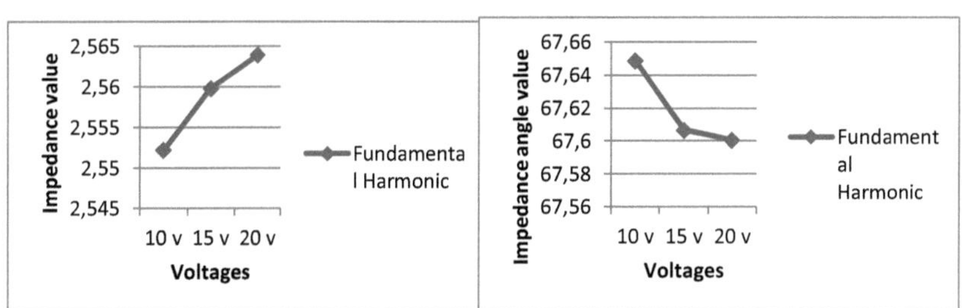

Figure 5.77 Plot of stator impedance at the 60 Hz fundamental harmonic and the value of the angles with waveform DST030.

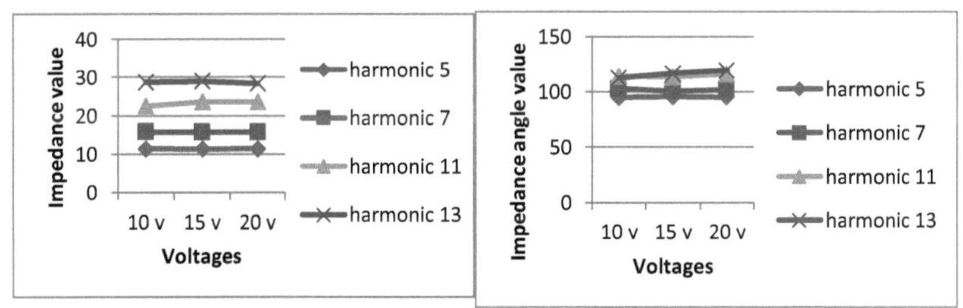

Figure 5.78 Plot of the stator impedance of the harmonics and the value of the angles with waveform DST030.

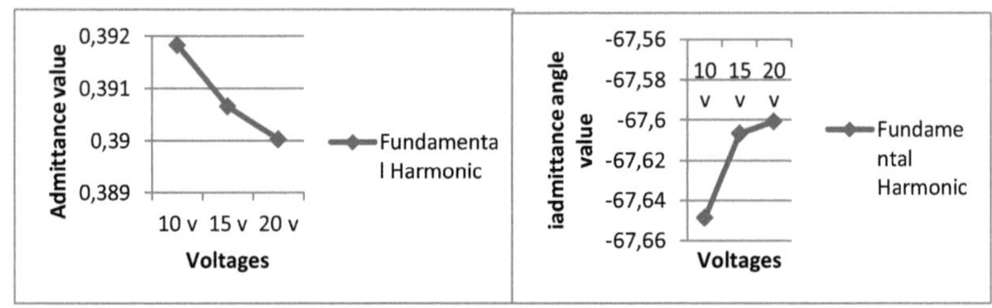

Figure 5.79 Plot of stator admittance at the 60 Hz fundamental harmonic and the value of the angles with waveform DST030.

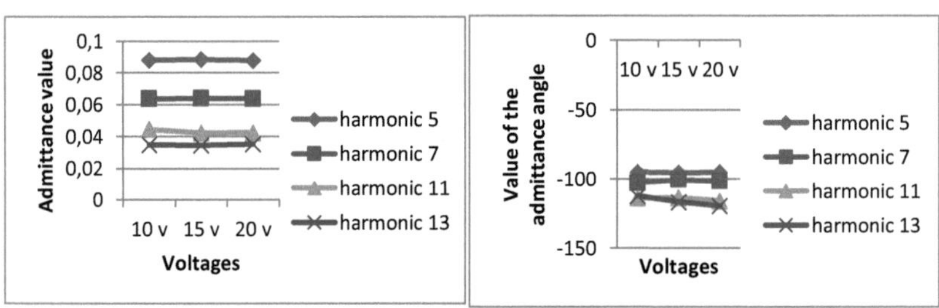

Figure 5.80 Plot of stator admittance at harmonics and angle value with waveform DST030.

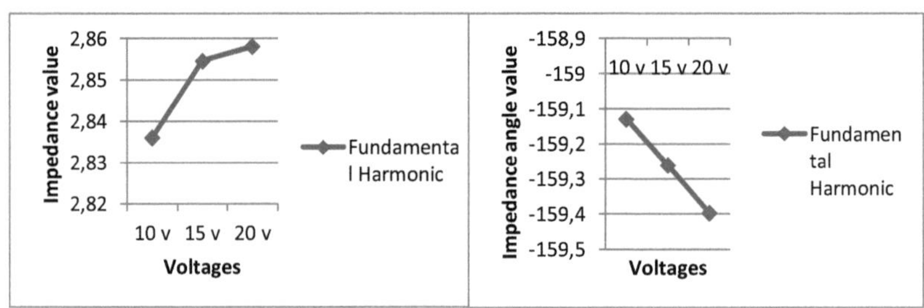

Figure 5.81 Plot of rotor impedance at the fundamental 60 Hz harmonic and the value of the angles with waveform DST030.

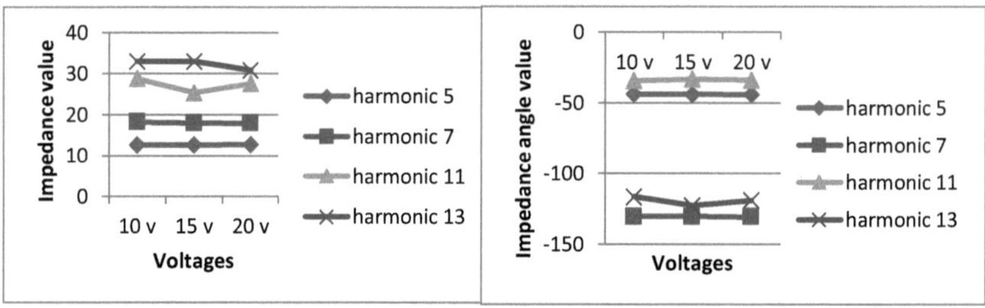

Figure 5.82 Plot of harmonic rotor impedance and angle value with waveform DST030.

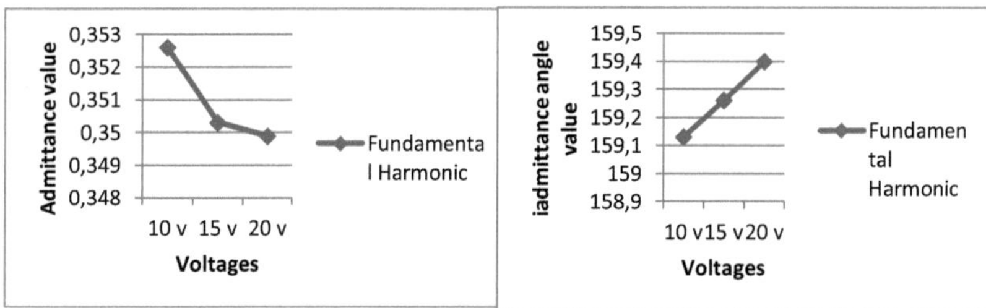

Figure 5.83 Plot of the rotor admittance at the fundamental 60 Hz harmonic and the value of the angles with waveform DST030.

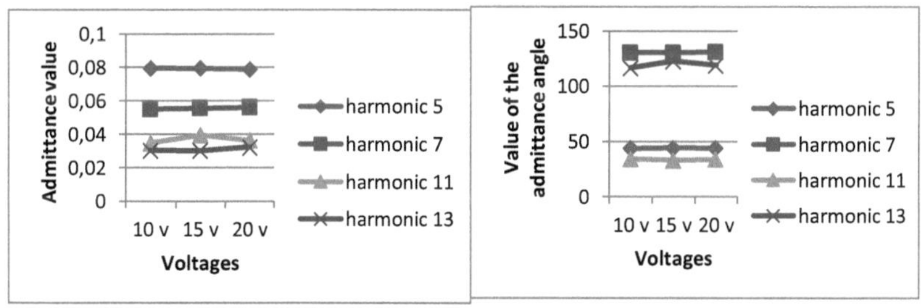

Figure 5.84 Plot of harmonic rotor admittance and angle value with waveform DST030.

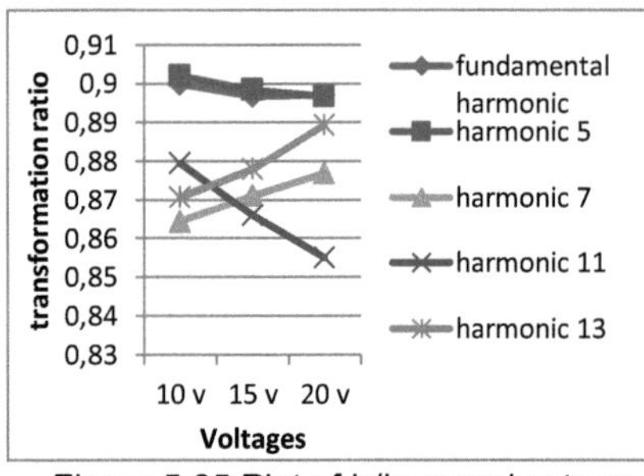

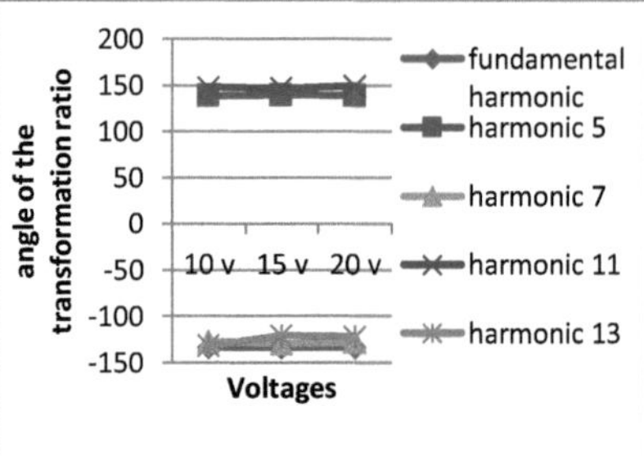


Figure 5.85 Plot of Ir/Ie complex transformer ratio with DST030 waveform and harmonic inductance at different voltages.

96

**CONCLUSIONS AND RECOMMENDATIONS**

At the end of the tests carried out on the electrical machine, the LabVIEW graphical comparisons of currents, voltages and the presence of harmonics induced by the source were observed in detail. All these variables are observed both in the stator and in the rotor, as well as the measurement of the frequencies and magnitudes comparing the different tests carried out.

In the frequency sweep test it can be seen how at different frequencies the values of the currents and voltages change both in the stator and in the rotor, the higher the frequency the voltage varies and the currents decrease both in the rotor and in the stator due to the increase in our impedance, the same happens with the angles of the voltages and currents, these are affected by a downward slide as the frequency r+jxl increases as it tends to go to the resistive part.

In the last test when harmonics are injected to the electrical machine, the test was performed at low voltage all this in order not to make the electrical machine rotate, all this demonstrative to perform the monitoring of the harmonics in the program made in LabVIEW, in more real tests the voltage would be increased so that the electrical machine would start and rotate and thus observe the real negative effects of the harmonics mentioned above.

The results of the residency project meet the objectives set at the beginning.

# BIBLIOGRAPHICAL REFERENCES


- 1) J. Chapman. "Electrical Machines. Fourth Edition. McGraw-Hill. MéxicoD.F.
- 2) http://www.monografias.com/trabajos82/operacion-paralelo-generadores-sincronos/operacion-paralelo-generadores-sincronos2.shtml
- 3) Electrical Machines, A. E. Fitzgerald, Fifth Edition.
- www.lem.com/
- http://mexico.ni.com/
- http://www.ni.com/labview/esa/?icid=HP_FG_es-MX_09240906_0617_png
- EL ABC de la instrumentación, Gilberto Enriquez Harper,LIMUSA, Mexico 2012.

# ANNEX

**Features of the *Chroma* Model 6590 Programmable AC Power Supply.**

| Model | 6560 | 6590 |
|---|---|---|
| Output Phase | 1 (parallel or series) | 1 or 3 selectable |
| **OUTPUT RATINGS** | | |
| **Power/phase** | 6000VA | 9000VA |
| **Voltage** | | |
| Range/Phase | 150V/300V(parallel) <br> 300V/500V(series) | 150V/300V |
| Accuracy | 0.2% reading + 0.2% of F.S | 0.2% reading + 0.2% of F.S |
| Resolution | 0.1V | 0.1V |
| Distortion | 1% | 1% |
| Line Regulation | 0.1% | 0.1% |
| Load Regulation | 0.2% (series) <br> 0.8 % (parallel) | 0.2% (3 phases) <br> 0.8% (1 phase) |
| Temp. coefficient | 0.02% per °C | 0.02% per °C |
| **Max. current/phase** | | |
| rms | 60A/30A/15A <br> (150V/300V/500V) | 30A/15A <br> (150V/300V) |
| peak | 180A/90A/45A(45-100Hz) <br> 150A/75A/38A(>100-1KHz) | 90A/45A(45-100Hz) <br> 75A/38A(>100-1KHz) |
| **Frequency** | | |
| Range | 45-1KHz | 45-1KHz |
| Accuracy | 0.15% | 0.15% |
| Resolution | 0.01Hz (45-99.9Hz) <br> 0.1Hz (100-999.9Hz) | 0.01Hz (45-99.9Hz) <br> 0.1Hz (100-999.9Hz) |
| **INPUT RATINGS** | | |
| Voltage Range | 190-250V, 3φ | 190-250V, 3φ |
| Frequency Range | 47-63Hz | 47-63Hz |
| Current | 23A Max. /Phase | 23A Max. /Phase |
| Power Factor | 0.98 Min. under full load | 0.98 Min. under full load |
| **MEASUREMENT** | | |
| **Voltage/Phase** | | |
| Range | 0-150V/0-300V | 0-150V/0-300V |
| Accuracy (rms) | 0.1%F.S.+0.25% | 0.1%F.S.+0.25% |
| Resolution | 0.1V | 0.1V |
| **Current/phase** | | |

| Range (peak) | 0-280A | 0-140A |
| --- | --- | --- |
| Accuracy (rms) | 0.1%F.S.+0.4% | 0.1%F.S.+0.4% |
| Accuracy (peak) | 0.2%F.S.+0.4% | 0.2%F.S.+0.4% |
| Resolution | 0.01A | 0.01A |
| **Power/phase** | | |
| Accuracy | 1%F.S. (CF<6) | 1%F.S. (CF<6) |
| Resolution | 0.01W | 0.01W |
| **Frequency** | | |
| Range | 45-1K Hz | 45-1K Hz |
| Accuracy | 0.01%+2count | 0.01%+2count |
| Resolution | 0.01Hz | 0.01Hz |
| **OTHERS** | | |
| Efficiency | 80% typ. | 80% typ. |
| Protection | OPP, OCP, OTP, FAN Fail | OPP, OCP, OTP, FAN Fail |
| Temperature | | |
| Operating | 0-40°C | 0-40°C |
| Storage | -40 to +85°C | -40 to +85°C |
| SAFETY | CE (include LVD and EMC requirement) | |

**Remarks**

*1: Maximum distortion for voltage varies from half to full range with linear load.

*2: Test with sine wave and remote sense.

# Source front panel

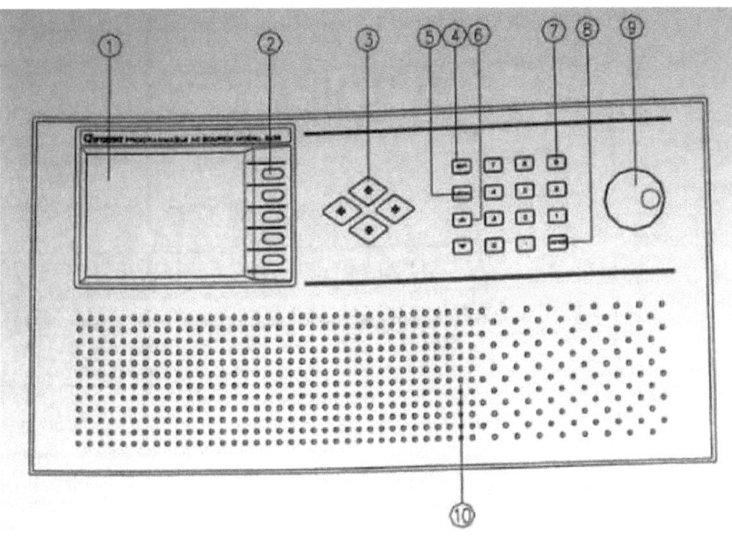

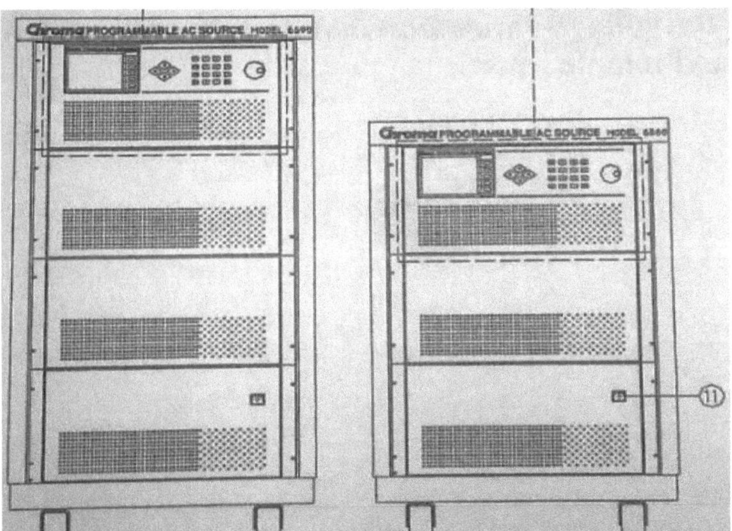

Front panel description

| Number | Symbol | Description |
| --- | --- | --- |
| 1 | | Display: The LCD displays the configuration, output settings and measurement results. |
| 2 | | Function keys: Five function keys are located on the right side of the display. each of them gives a command as indicated in the corresponding command block on the LCD display next to the function keys to guide the user unless the block is blank. Each valid function key denotes the command string lying on the symbol of a function key; for example, CONFIG represents the function key next to the command block .config. command to enter the configuration menu. |
| 3 | ↑ ← → ↓ | Moving cursor keys: these four keys are for moving the cursor to different directions, respectively. in normal mode (non-editing mode), pressing any of them will change the cursor position. in editing mode, pressing .. will move the digit cursor by digit for programming the parameter. |
| 4 | Edit | Edit command key: Press this key to enter or exit the edit mode during V or F programming. |
| 5 | Local | Local Command Key: Press this key to switch control of the AC sources from remote to local operation using the keypad on the front panel. |
| 6 | ▲ and ▼ | Up and down keys: Press ▲ to search upwards for the desired programming parameters. Press ▼ to search downwards. |
| 7 | 0 to 9 and . | Numeric and decimal keys: you can program numeric data by pressing the numeric and decimal keys. |
| 8 | ENTER | Enter password: press this key to confirm the parameter settings. |

| | | |
|---|---|---|
| **9** | | RPG: You can enter programming data or options by turning the RPG to the desired ones. |
| **10** | | Airflow inlets: Cold airflow enters the AC source through these holes. |
| **11** | | Main power switch: This is for turning the device on or off. |

# Rear panel of the source

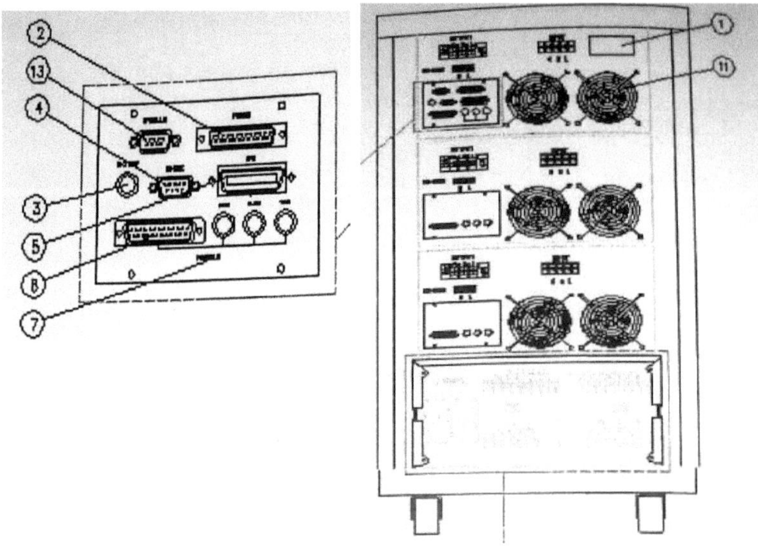

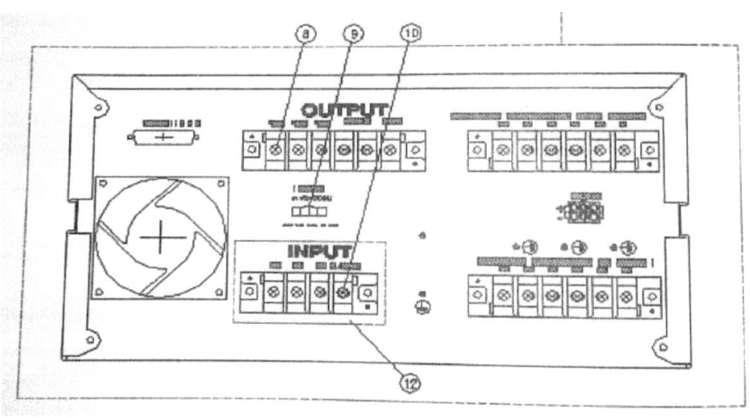

Description of the rear panel

| Number | Name | Description |
| --- | --- | --- |
| 1 | Label | The label contains the model number, serial number of the AC source and its input ratings. |
| 2 | Printer port. | The 25-pin female connector, type D is for transmitting the measurement data to a parallel printer. |
| 3 | Ext. Ref. | BNC connector that controls the amplitude of the waveform of external device inputs. |
| 4 | RS-232C. | The 9-pin, D-type female connector is for control of transfers to and from the remote PC for remote operation. |
| 5 | GPIB connector | The GPIB connector is for connecting a bus to remotely operate the AC source. |
| 6 | System interface: I/O system | The 25-pin, D-type male connector is for parallel, three-phase data communication. |
| 7 | System interface | These three BNC signals: SYNC, CLOCK, VREF are for three-phase parallel synchronisation. |
| 8 | Output connector. | This connector delivers power to the charging device. |
| 9 | Connector Remote sensing | It senses directly on the load terminals to eliminate any voltage drop on the connection cable. Be sure to connect the remote connector terminal direction "L" to the load terminal "L" and "N" to the load "N". Reversed polarity is not allowed. |
| 10 | Power connector | This input is for supplying the source with three-phase voltage. |
| 11 | Cooling fan. | Fan speed automatically increases or decreases as temperature rises or falls. |
| 12 | Security cover | This is to protect the user from being electrocuted. |
| 13 | TTL I/O. | The 9-pin, D-type male connector transfers trigger control signals  on, remote inhibit, transient and AC ON. |

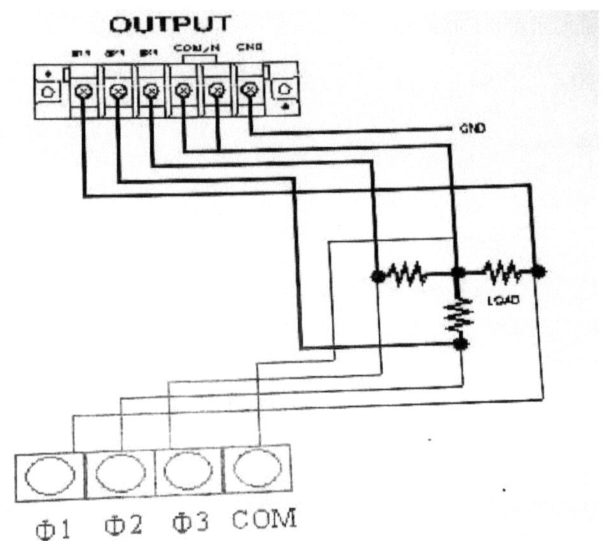


Connection for three-phase output.

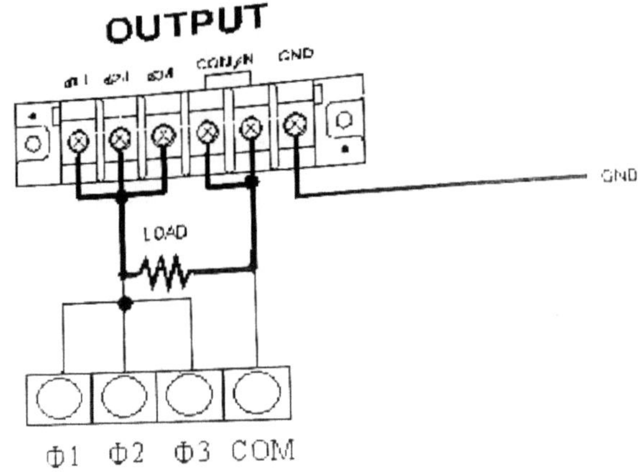


Connection for single-phase output.

More
Books!

info@omniscriptum.com
www.omniscriptum.com

OMNIScriptum